T0262730

Eminent Topics in Conservation Biology

Eminent Topics in Conservation Biology

Edited by **Neil Griffin**

New York

Published by Callisto Reference,
106 Park Avenue, Suite 200,
New York, NY 10016, USA
www.callistoreference.com

Eminent Topics in Conservation Biology
Edited by Neil Griffin

International Standard Book Number: 978-1-63239-169-8 (Hardback)

Printed in the United States of America.

Contents

	Preface	VII
Chapter 1	**Biochemical Diversity of Wild Carob Tree Populations and Its Economic Value** S. Naghmouchi, M. L. Khouja, A. Khaldi, M. N. Rejeb, S. Zgoulli, P. Thonart and M. Boussaid	1
Chapter 2	**Protected Areas in Selected Arab Countries of the Levant Region (Syria, Lebanon & Jordan): An Evaluation of Management and Recommendations for Improvement** Brandon P. Anthony and Diane A. Matar	17
Chapter 3	**Introgression and Long-Term Naturalization of Archaeophytes into Native Plants – Underestimated Risk of Hybrids** Hiroyuki Iketani and Hironori Katayama	43
Chapter 4	**Amazonian Manatee Urinalysis: Conservation Applications** Tatyanna Mariúcha de Araújo Pantoja, Fernando César Weber Rosas, Vera Maria Ferreira Da Silva and Ângela Maria Fernandes Dos Santos	57
Chapter 5	**Global Efforts to Bridge Religion and Conservation: Are They Really Working?** Stephen M. Awoyemi, Amy Gambrill, Alison Ormsby and Dhaval Vyas	81
Chapter 6	**Managing Population Sex Ratios in Conservation Practice: How and Why?** Claus Wedekind	95
	Permissions	
	List of Contributors	

Preface

Conservation biology is an essential element for sustaining the biodiversity of this planet. Conservation biology is also known as "crisis discipline." In today's scenario, the world is undergoing some rapid transformation, and science instructs us about research, management practices, technologies and policies that can help protect the environment's naturally-occurring biological diversity. Certain chapters of this book present well-researched analysis dealing with guarded regions (Middle East), safeguarding biochemical and genetic diversity of carob tree, wild pear, shaping the health status of Amazon manatee, disturbing sex-ratios to benefit the natural world, and minimizing the gap between region and protection. This book approaches various threats to biological diversity from different point of views. The author offers room for reflection on the meaning and value of the word 'natural' on a planet now, tremendously, dominated by people.

This book has been the outcome of endless efforts put in by authors and researchers on various issues and topics within the field. The book is a comprehensive collection of significant researches that are addressed in a variety of chapters. It will surely enhance the knowledge of the field among readers across the globe.

It is indeed an immense pleasure to thank our researchers and authors for their efforts to submit their piece of writing before the deadlines. Finally in the end, I would like to thank my family and colleagues who have been a great source of inspiration and support.

Editor

1

Biochemical Diversity of Wild Carob Tree Populations and Its Economic Value

S. Naghmouchi[1*], M. L. Khouja[1], A. Khaldi[1], M. N. Rejeb[1],
S. Zgoulli[2], P. Thonart[2] and M. Boussaid[3]
[1]*Institute of Research in Rural Engineering, Water and Forestry (INRGREF), Tunisia*
[2]*Wallon Centre of Industrial Biology (CWBI). University of Liege Sart Tilman, Liège,*
[3]*National Institute of Applied Science to Technology (INSAT Tunis),*
Centre Urbain Nord, Tunisia
[1,3]*Tunis*
[2]*Belgique*

1. Introduction

Carob tree *Ceratonia siliqua* L. has been included in a national list of priority forest genetic resources for conservation and management in Tunisia (Bouzouita et al., 2007). The reduction of the biochemical diversity of this species in this region is a real risk. The species belongs to the Cesalpinaceae sub-family of the family Leguminoseae (syn Fabaceae).

It is an evergreen tree that is widespread in the Mediterranean basin. The species is mainly used for animal feeding and pharmaceutical and food industries (Yousif & Alghzawi, 2000; Tous et al., 2006; Silanikove et al., 2006).

Carob pods provide two important products: carob kernels from which carob or locust bean gum (LBG) is extracted, and carob kibbles or the remaining pulp obtained after the removal of the kernels. This can be used directly in animal and human nutrition (Mhaisen, 1991; Silanikove et al., 2006) or as a raw material for industrial processing (Carlson, 1986).

Pods are characterized by a high sugar content with about 75% of those sugars are sucrose (Biner et al., 2007). Carob kernels (10-20% of the fruit weight) contain principally galactomannans (Egli, 1969; McCleary & Matheson, 1976). The Locust bean gum is the ground endosperm of seeds; owing to its remarkable water-binding properties, it is widely used to improve food texture (Imeson, 1997). Gum-aqueous solutions have a high viscosity and are used as a substitute for pectins, agar or other mucilagenous substances (Fulgancio et al., 1982). The purification of polysaccharides improves this situation; unacceptable flavours of the crude gums are removed and the purified gums give clear and more stable solutions due to the elimination of impurities and endogenous enzymes. There are several methods for purificated crude galactomannan samples. Precipitation with ethanol has been largely used (Doublier, 1975). Purification by methanol (Rafique & Smith, 1950) and by copper

---

(McCleary & Matheson, 1976) or barium complexes (Kapoor, 1972) has also been used; isopropanol is the best method for industrial processes (Bouzouita et al., 2007).

The aim of this study is to assess the variation of physical and chemical traits (moisture, ash, pH, protein, acid-insoluble matter, fibres, total phenols, sugars contents and mannose/galactose ratio of crude samples) among Tunisian natural populations. The rheological behaviour of the purified and crude bean gum is also assessed. Our study is an extension of those performed to assess the genetic diversity of Tunisian carob populations (Afif et al., 2006) (via izozymes and molecular markers) in order to elaborate conservation and product improvement strategies.

2. Methods

2.1 Analysed populations and sampling

Twelve Tunisian wild carob populations were analysed (Figure 1).

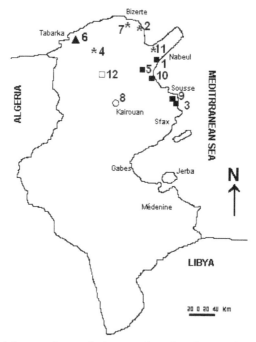

Fig. 1. Location of Tunisian carob populations analysed in this study.
Numbers indicate populations: **1**: Hammamet; **2**: Gharelmelh; **3**: Sayada; **4**: Slouguia; **5**: Jradou; **6**: Ainessobh; **7**: Menzel Bourguiba; **8**: Chbika; **9**: Khnis; **10**: Enfidha; **11**: Slimen; **12**: Bargou.
▲: Low humid; *: Sub humid; □: Upper Semi arid; ■: Low Semi arid; ○: Upper arid.

Populations belong to the upper arid, low semi arid, sub humid and low humid bioclimates according to Emberger's (1966) pluviothermic coefficient (Q2) 17. Pods were collected during the period of July-September 2009. Five to 10 trees per population sampled at random were considered, and 25 randomly selected mature pods per tree were analysed.

2.2 Experimental methods

2.2.1 Chemical analysis

Mature pods collected in each population were air dried for 4 weeks and kept at 4°C before analyses. Pod pulp and kernels were ground in particles less than 500μm and 180μm respectively. The resulting powder: pulp (p) and non purified gum (g) powder was analysed for moisture (Mp, Mg), ash contents (Ap, Ag) and pH (pHp, pHg) according to Food Chemical Codex. The total nitrogen was evaluated by the Kjeldahl's method. Crude protein (Prp, Prg) contents were calculated by multiplying the nitrogen content by 6.25 (Anderson, 1986). Total phenols (Php, Phg) in pulp and non purified locust bean gum (LBG) were determined by Folin-Ciacolteu's method. Neutral Detergent Fibre (NDFp, NDFg) and Acid Detergent Fibre (ADFp, NDFg) were estimated according to AOAC method.

D-glucose (Gp), saccharose (Sp) and D-fructose (Fp) in pods pulp were determined by enzymatic kits and the determination of Optic Density at 750nm with spectrophotometer (UV/Vis).

To assess mannose and galactose ratios (M/G), twenty millilitres of H_2SO_4 (1M) were added to 400 mg of carob bean gum (purified and crude gum). After boiling, the mixture was cooled and treated with $BaCO_3$ and adjusted to pH 7, filtered and evaporated (30°C and 50°C) to obtain crystalline residue or syrup. The precipitate was dissolved in 2ml of desionised water and analysed by HPLC using an Agilent HP chromatograph, series 1100 with waters differential refraction detector and equipped with a Pb^{2+} column at 80°C.

The contents of galactomannan were deduced from a calibration curve obtained from five synthetic mixtures of β-D-galactose and β-D-mannose in known proportions.

2.2.2 Purification of the gum

The required amount of the powdered gum was gradually added to highly stirred distilled water. The precipitation is initiated at 13g.l-1 concentrated aqueous solutions. The solution was moderately stirred at different temperatures (from 25°C to 80°C) for 90 minutes. The resulting solution was allowed to cool and then centrifuged at 21875 g at 19°C. The obtained visquous supernatant corresponding to the crude LBG solution was separated for its use in the purification process. The solubilized galactomannan was precipitated from crude LBG solution by pouring into a two-volume excess of isopropanol. The precipitate was collected by filtration and lyophilisated for 5h at 40°C, 3 h at 5°C and for 3 days at 20°C, respectively.

2.2.3 Viscosity determination

The rheological properties were assessed by the measurment of the viscosity of the crude and purified LBG solutions, using an Ares Advanced Rheometric Scientific, with a coaxial cylindrical system rheometer which works under both dynamic and static regimes. The samples were sheared in the gap between the fixed inner cylinder (U = 10.1 mm, h = 61.4 mm) and the rotated mobile outer cylinder (U = 11.5 mm) and sample volume is 12.0cm3. To prepare 2% and 1% aqueous solutions of crude and purified gum, we add to 360ml distilled water, 8g and 4g respectively of crude and purified gum. The solution was mixed at 25°C for 15 minutes and at 90°C for 15 minutes and then remained at ambient temperature. The viscosity was measured at different shear rates.

2.3 Statistical analysis

The heterogeneity for each measured parameter among populations was tested using variance analysis with one effect (population or bioclimate effect) using the program SAS, procedure ANOVA (SAS, 1990). Averages of parameters were compared using Duncan's multiple range test (at P<0.05).

A principal component analysis (PCA) based on all considered parameters was performed to assess the divergence among populations, using the program MVSP version 3.1 (Kovach, 1999). Cluster analysis was performed on the matrix of Euclidean distances for the estimation of the physico-chemical variability distribution among populations and ecological groups. The divergence among populations and their locations (bioclimatic zones) was assessed by Euclidean distances calculated between pairs of populations and by the construction of a dendrogram generated from these distances using the UPGMA (Unweighted Pair Group Method with Arithmetic Averages) method. The distances matrix determined by cluster analysis was correlated with those of geographic distances.

3. Results

3.1 Variation of pulp parameters

Averages of the analysed parameters are presented in table 1. A significant variation (P<0.001) was observed among populations for all examined parameters.

Pods had low moisture (158 g kg $^{-1}$) and values of Mp varied from 93 g kg $^{-1}$ (population 9) to 234 g kg $^{-1}$ (population 5). The crude protein contents ranged from 2.8 g kg $^{-1}$ (population 4) to 7.1 g kg $^{-1}$ (population 5). Acid and Neutral Detergent Fibres contents showed high proportions, they ranged from 2.4 g kg $^{-1}$ to 3.7 g kg $^{-1}$ and from 2.8 to 3.9 g kg $^{-1}$, respectively. Population 8 from the low semi-arid bioclimate showed the lowest values of NDF and ADF, while the upper arid population 7 and the low semi-arid population 11 showed the highest ones.

The average of Ash equalled 0.21 g kg $^{-1}$. Populations were subdivided into two groups without apparent correlation to geographic or bioclimatic locations of populations. The first group showing low values (0.15 to 0.18 g kg $^{-1}$) included populations 2, 3, 7 and 8 and the second group included the other populations which were characterized by high Ash values.

The amounts of phenols varied highly among populations. Populations 5 (2.3 g kg $^{-1}$), 4 (2.1 g kg $^{-1}$), 12 (2.1 g kg $^{-1}$) and 11 (2.0 g kg $^{-1}$) had the highest amount of phenols while the group of populations 7, 1, 9, 3 and 8 showed the lowest values.

Regarding sugar content, it is obvious that carob pulp contained high levels of sugar: 2.8 g kg $^{-1}$ sucrose, 0.36 g kg $^{-1}$ D-glucose and 0.46 g kg $^{-1}$ D-fructose. The level variation of sugars contents among populations was higher than that for the other constituents. In fact, 8 groups of populations were observed.

3.2 Variation of locust bean gum parameters

The crude locust bean gum has an average of 0.27 g kg $^{-1}$ Ash (Ag). A high variation among populations was revealed (table 2), with populations 11, 6 and 12 showing the highest values (0.28 to 0.31 g kg $^{-1}$).

Population code	Mp		AIMp		Ap		Prp		php		Fibre			Sugar				
											ADFp		NDFp		Gp		Sp		Fp	
1	109	e	2.0	bc	0.24	a	3.5	c	1.1	e	3.1	c	3.8	b	0.44	c	4.4	a	0.46	e
2	187	cd	2.1	ab	0.18	b	4.4	b	1.9	cd	2.8	e	3.4	e	0.47	a	2.7	h	0.55	d
3	94	e	1.6	de	0.17	b	5.1	b	0.9	e	3.0	cd	3.5	d	0.37	e	2.9	f	0.38	i
4	212	b	1.6	e	0.22	a	2.8	c	2.1	b	2.8	e	3.5	c	0.35	f	3.3	b	0.66	a
5	234	a	1.6	de	0.23	a	7.1	a	2.3	a	3.5	b	3.6	c	0.44	c	3.0	d	0.45	f
6	171	d	1.9	bcd	0.25	a	6.5	a	1.7	d	2.6	f	3.6	c	0.13	c	3.2	c	0.28	k
7	108	e	1.7	de	0.16	b	4.9	b	1.1	e	2.5	g	3.0	i	0.24	i	1.9	l	0.56	b
8	108	e	1.8	cde	0.15	b	3.4	c	1.1	e	2.4	g	2.8	j	0.46	b	2.3	j	0.44	f
9	93	e	1.8	cde	0.22	a	6.4	a	0.9	e	3.1	cd	3.2	g	0.34	g	2.8	g	0.36	j
10	174	d	2.0	b	0.21	a	3.5	c	1.7	d	3.7	a	3.1	h	0.31	h	3.0	e	0.39	h
11	195	bc	1.9	bcd	0.23	a	3.7	c	2.0	bc	3.0	d	3.3	f	0.37	e	2.5	i	0.54	c
12	209	b	2.3	a	0.22	a	4.6	b	2.1	b	3.4	b	3.9	a	0.42	d	2.3	k	0.43	g
Over all popuations	15.8***		1.9***		0.21***		4.7***		1.6***		3***		3.4***		0.36***		2.9***		0.46***	

Table 1. Average of the pulp measured parameters (g kg^{-1}) for populations.
Values with different letters in the same trial column differ significantly (P<0.05), *** Highly significant at p<0.001.
Mp: Moisture, pHp: pH, AIMp: Acid insoluble matter, Ap: Ash, Prp: Protein, ADFp: Acid Detergent Fibre, NDFp: Neutral Detergent Fibre, php: Total phenols, Gp: D -Glucose, Sp: Sucrose, Fp: D-Fructose.

The content of crude bean gum protein (Prg) varied from 1.6 g kg^{-1} (population 3) to 1.9 g kg^{-1} (population 5) with an average of 1.7 g kg^{-1}, and an average of 1.1 g kg^{-1} acid insoluble matter (AIM)g.

Six and eight groups of population were distinguished respectively for crude bean gum ADF and NDF. The ADFg values (1 g kg^{-1}) varied between 1.3 g kg^{-1} (population 3) and 0.8 g kg^{-1} (population 7), while those NDFg (3.5 g kg^{-1}) ranged from 3.9 g kg^{-1} (population 8) and 2.8 g kg^{-1} (population 3).

The average of moisture content of crude LBG (80 g kg^{-1}) was lower than that of carob pulp, the pH pulp value (5.2) was less than the pH of LBG (6.0). Both pulp and locust bean gum were remarkably rich in NDF (3.4 and 3.5 g kg^{-1} respectively).

The mannose/galactose ratios (M/Gg) varied among gum samples (Table 2). Crude LBG samples showed higher M/Gg ratios. The highest ratio (M/Gg= 4.2) was observed for

Population code	Mg		AIMg		Ag		Prg		phg		Fibre				Yg		M/G ratio			
											ADFg		NDFg				M/Gg		M/Gpg	
1	90	a	0.9	c	0.23	f	1.8	ab	2.1	b	0.9	g	3.4	C	389.9	abc	4.2	a	3.9	abc
2	79	cd	1.1	b	0.27	c	1.7	ab	1.3	e	1	d	3.5	f	388.8	abc	3.3	de	3.9	abc
3	78	d	1.2	a	0.27	c	1.6	c	1.2	e	1.3	a	2.8	f	340.2	bc	4	b	3.4	bc
4	78	d	1.1	ab	0.25	d	1.7	bc	2.1	b	1.1	b	3.6	g	337.9	bc	4	b	3.4	bc
5	78	d	0.8	d	0.25	de	1.9	a	1.9	c	0.8	h	3.2	e	435.3	a	4	b	4.4	a
6	83	c	1.1	b	0.29	b	1.8	a	1.7	d	0.9	e	3.8	f	412.0	ab	3.2	de	4.1	ab
7	87	b	1.1	ab	0.24	e	1.7	ab	1.2	e	0.8	h	3.5	b	453.7	a	3.5	c	4.5	a
8	71	e	1.1	ab	0.26	c	1.8	a	2.3	a	1.1	b	3.9	b	382.7	abc	3.9	b	3.8	abc
9	81	cd	0.9	c	0.26	c	1.6	c	1.9	c	0.9	ef	3.1	d	406.0	ab	3.3	d	4.1	ab
10	79	cd	1.1	b	0.27	c	1.7	bc	1.6	d	1.1	c	3.3	c	323.5	c	3.5	c	3.2	c
11	81	cd	1.1	ab	0.28	b	1.7	bc	1.5	d	0.9	fg	3.6	a	385.3	abc	3.9	b	3.9	abc
12	80	cd	1.1	b	0.31	a	1.7	abc	1.4	e	1.1	b	3	h	344.7	bc	3.1	e	3.5	bc
Over all populations	80***		1.1***		0.27***		1.7***		1.7***		1***		3.5***		383.4***		3.67***		3.83***	

Table 2. Average of the locust bean gum measured parameters (g kg $^{-1}$) for populations Values followed by different letters in the same trial column differed significantly (P<0.05), *** Highly significant at p<0.001.
Mg: Moisture, pHg: pH, AIMg: Acid insoluble matter, Ag: Ash, Prg: Protein, ADFg: Acid Detergent Fibre, NDFg: Neutral Detergent Fibre, phg: Total phenols, Yg: Yield (% total gum. dry weight basis), Mannose/Galactose ratio of crude LBG: M/Gg, Mannose/Galactose ratio of purified LBG: M/Gpg.

population 1 (from the low humid climate). Populations 3, 4 and 5 (from sub humid) with 8 and 11 (from the low semi arid) showed also high M/Gg ratios but did not exhibit significant differences. Populations 7 and 10 had intermediate values. Populations 2, 6, 9 and 12 showed the lowest M/Gg ratios.

Purified LBG showed also higher M/Gpg ratio (3.83). Low and non significant differences among populations were observed, the population 10 showed the lowest value (3.2) and the populations 7 and 5 showed the highest values (4.5 and 4.4 respectively).

The level of ANOVA, which considered the twenty populations as five groups (according to bioclimate) indicated that all constituents (except protein Prg) differed significantly ($P < 0.05$). Variation observed between populations belonging to the same bioclimatic zone was not significant for all characters.

Populations from the sub-humid bioclimate had shown significant differences in all characters except ADFp pulp and ADFg gum contents (p=0.89 and p=0.26 respectively), gum moisture (p=0.38) and Mannose/Galactose ratio of pure gum (p=0.2). Within the Lower semi arid ecological group, the major parameters (13/24 parameters) were significantly different at p<0.05.

The principal component analysis showed that the first three axes explain 80.72% of the total variation. A specific meaning could be variables as follows:

- The first axis (48.35%) is correlated to Acid insoluble matter (AIMp), Neutral Detergent fibre (NDFp, NDFg), yield of gum (Yg) and Mannose/Galactose (M/Gp) of purified gum.
- The second axis explained 20% of total inertia and correlated to pH and phenols of pulp and gum (pHp, pHg, Php, Phg), glucose pulp content (Gp), acid insoluble matter gum amount (AIMg) and protein locust bean gum content (Prg).
- The third component, instead, explained 12.37% of total variance. Loaded variables were moisture, ash and acid detergent fibre contents of pulp and gum (Mp, Mg, Ap, Ag, ADFp, ADFg), saccharose (Sp), fructose (Fp), protein (Prp), and Mannose/Galactose (M/Gg) of crude gum and viscosity (Vg, Vp) of crude and purified gum

The plot according to axes 1-2 and axes 1-3 (Figure 2 and 3) showed a high dispersal of populations.

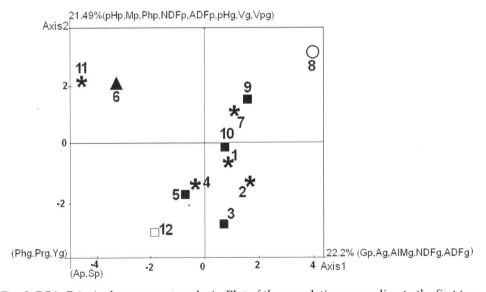

Fig. 2. PCA: Principal component analysis. Plot of the populations according to the first two components. Codes indicate populations. 1: Hammamet; 2: Gharelmelh; 3: Sayada; 4: Slouguia; 5: Jradou; 6: Ainessobh; 7: Menzel Bourguiba; 8: Chbika; 9: Khnis; 10: Enfidha; 11: Slimen; 12: Bargou.
▲: Low humid; *: Sub humid; ■: Low Semi arid; □: Upper Semi arid; ○: Upper arid.

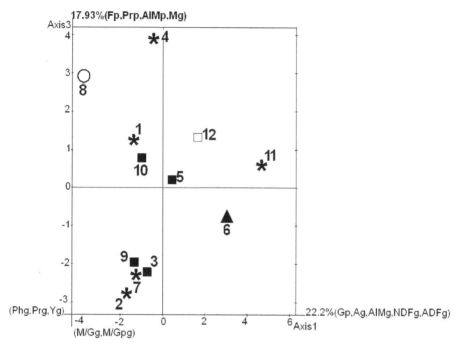

Fig. 3. PCA: Principal component analysis. Plot of the populations according to the first and the third components. Codes indicate populations. 1: Hammamet; 2: Gharelmelh; 3: Sayada; 4: Slouguia; 5: Jradou; 6: Ainessobh; 7: Menzel Bourguiba; 8: Chbika; 9: Khnis; 10: Enfidha; 11: Slimen; 12: Bargou.
▲: Low humid; *: Sub humid; ■: Low Semi arid; □: Upper Semi arid; ○: Upper arid.

According to the axis 1, two major groups can be distinguished: the first includes populations from Low semi arid and populations from Upper semi arid zones, populations 7 and 6 from (the same bioclimate) were separated according to axis 2; the second on negative side of axis 1 is constituted by populations from Sub humid and upper arid zones, population 12 is well isolated from the others.

The dendrogram constructed using Euclidean distances (Fig. 4) showed four groups of populations:

-   The first group is constituted by two populations (7 and 9) respectively from Upper and Low semi arid bioclimates. This group is located in the lower-right of the graph (Fig. 2) with positive values of the second and third components.
-   The second group included 8 populations belonging to all bioclimates, two aggregates can be distinguished in this group; the first aggregate is made up of five populations (2, 3, 4, 5 and 6), all the populations clustered together according to their geographic proximity (Sub-humid bioclimate) except for the population 6 that belongs to the Upper semi arid bioclimate, the second aggregate included three populations (1, 10 and 11). In this aggregate, the population 1, belonging to the Lower humid bioclimate, was separated from populations 10 and 11 that belongs to the Low semi arid bioclimate.

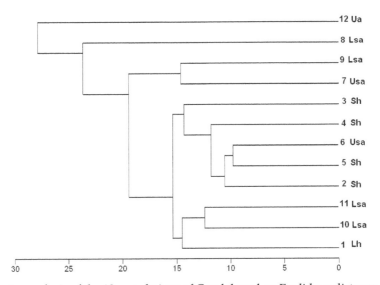

Fig. 4. Cluster analysis of the 12 populations of Carob based on Euclidean distance.
Codes indicate populations. 1: Ainessobh; 2: Hammamet; 3: Gharelmelh; 4: Menzel
Bourguiba; 5: Slimen; 6: Slouguia; 7: Bargou; 8: Sayada; 9: Jradou; 10: Khnis; 11: Enfidha; 12:
Chbika.
Bioclimatic zones: Lh: Low humid; Sh: Sub humid; Lsa: Low semi arid; Usa: Upper semi
arid; Ua: Upper arid.

Populations 12 (from the Upper arid) was alone considered as group. Contrary to
population 8, population 12 was located in the negative part of axis 3 with high and
negative values of the second and third components (Figure3).

The plot of the Principal Component Analysis (PCA) of the *Ceratonia siliqua* individuals
sampled showed that the first three axes described 79.7% of the total variation; The first,
second and third axis explained respectively 52.22%, 16.36% and 11.11% of total inertia (Fig.
5 and 6).

Euclidean distances (D) between populations ranged from 9.78 to 41.78. Mean D value for
all population was 20.97. The highest D-value was observed between populations 8 and 12
belonging respectively to the lower semi-arid and the upper arid bioclimatic zones and 139
km distant from each other. The lowest genetic distance (9.78) was recorded between
populations 5 and 6 which are geographically close (10 km). However the correlation
between geographic and Euclidean distance is not significant for all carob populations
($r^2 = 0.033$, p = 0.168) (Fig.5).

3.3 Yields of the gum

The extraction and purification processings were used to obtain the purified
galactomannans with efficiency varying from 323.5 g kg $^{-1}$ (population 10) to 453.7 g kg $^{-1}$
(population 7). The average value was 383.4 g kg $^{-1}$. Populations 5, 2, 6, 8, 9 and 1 presented
high yields (from 382.7 to 435.3 g kg $^{-1}$) but the values did not differ significantly.

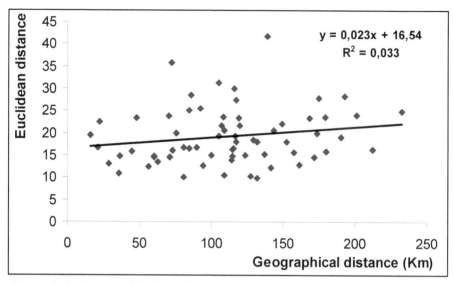

Fig. 5. Correlation between Euclidean distances and geographical distances (km) between all pair of populations.

3.4 Rheological characteristics

The flow curves of crude and purified LBG samples reported in figures 8 and 9 showed that all solutions had a shear-thinning behaviour showing that apparent viscosity decreased when the shear rate increased. The apparent viscosity at 6s-1 varied significantly from 0.04 to 4.78 Pa.s for crude LBG and from 0.07 to 4.83 Pa.s for purified LBG (Table 3), according to populations (P<0.001).

Population code	Crude Locust Bean Gum		Purified Locust Bean Gum	
1	4.37	b	4.6	a
2	1.55	f	1.8	d
3	1.65	f	1.78	d
4	2.95	d	3	c
5	4.78	a	4.82	a
6	0.04	h	0.07	f
7	1.25	f	1.45	d
8	2.17	e	2.7	c
9	0.81	g	0.9	e
10	3.5	c	3.8	b
11	3.42	h	4	f
12	1.45	f	1.72	d
Over all populations	2.33***		2.55***	

Table 3. Apparent viscosity (Pa.s) of crude and purified locust bean gum for the analysed populations. Values with different letters in the same trial column differ significantly (P<0.05) *** Highly significant at p<0.001.

The best rheological properties were observed for populations 5 (4.78 Pa.s and 4.83 Pa.s respectively for crude and purified LBG) and 1 (4.38 Pa.s and 4.6 Pa.s respectively for crude and purified LBG). The lowest apparent viscosity at 6s-1 was obtained for population 6 with 0.04 and 0.07 Pa.s for crude and purified LBG respectively. Figures 6 and 7 show that purified LBG solutions had a higher shear-thinning (response of a fluid's viscosity to a shearing stress, that is, a force tending to make part of the fluid slide past another part) than the crude LBG solutions.

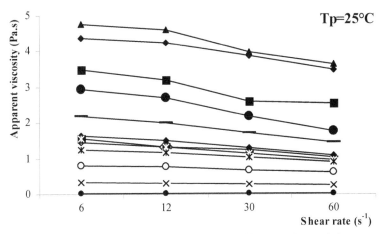

Fig. 6. Apparent viscosity of crude locust bean gum at different values of shear rate for samples from all populations.

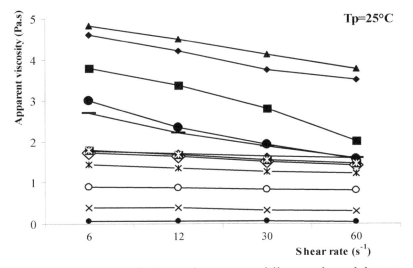

Fig. 7. Apparent viscosity of purified locust bean gum at different values of shear rate for samples from all populations.

Each symbol indicates a population:

4. Discussion

4.1 Biochemical properties

The moisture values of carob pulp from Tunisian natural populations were higher than those previously reported for carob collected from Portugal, Spain and Jordan (Blenford, 1979; Brandt, 1984; Albanell et al., 1991). Our results confirmed that, as reported by Lopes da Silva & Gonçalves (1990), the differences may be attributed to the difficulties experienced during the milling of carob pods. In our experiments, the level of variation differed according to population and bioclimate. The pH of carob pulp was lower than that reported in the literature, due probably to caramelisation during the powdering process and to the formation of by-products and other intermediates such as pyruvic acid (Lee et al., 1990).

Ash and protein contents showed significant differences among populations, yet the obtained values were similar to the FAO standard ones. The ash, phenol and protein values of carob pulp are lower than those of kernels farina or LBG. These results are consistent with earlier findings (Blenford, 1979; Youssif & Alghzawi, 2000). The low protein content may support the assumption that carob could be considered a natural healthy food (Bengoechea et al., 2008). On the other hand, carob pulp is free from the two anti-nutrients caffeine and theobromine (Craig and Nguyen, 1984) and has higher rates of crude fibre. Absence of caffeine might be considered an advantage, from a nutritional point of view, qualifying pulp as a good replacer or extender for cocoa powder in many food items. The moderate amount of phenols in pulp favored for its use as ingredient in animal feed (Tamir & Nachtomi, 1973) and dietetic products, or as a cocoa and sugar replacer in human food free from caffeine or theobromine. Individuals from populations such as populations 11, 7 and 8 could be selected to promote low phenol varieties.

Sucrose, fructose and glucose contents of pulp were high and found to differ significantly among populations. Sucrose is the dominant sugar found in carob. Our results are similar than those found by Biner et al. (2007).

The high yield or gum percentage is one of the important required parameters for carob industry, because locust bean gum was the first galactomannan used both industrially (paper, textile, pharmaceutical, cosmetic and other indus-tries) and in food products (ice cream and other preparations). In our experiments, the yield oscillated between 323.5 g kg $^{-1}$ and 453.7 g kg $^{-1}$ and showed significant differences between populations. Bargou followed by Slimen, Slouguia and Jradou were the populations with higher amounts of gum. In the literature, the yields were approximately between 290 g kg $^{-1}$ and 460 g kg $^{-1}$ (Andrade et al., 1999).

The apparent viscosity varied with shear rate. The behaviour is pseudoplastic at every shear rate, in agreement with other studies (Garcia & Casas, 1992). The rheological properties of locust bean gum did not depend only upon protein, but the most important factor was the galactomannan. The purification of the samples of crude locust gum eliminated practically all fat and fibre; the ash and protein contents may drastically be reduced (Bouzouita et al.,

2007). This diversity in biochemical composition of pulp and kernel and the physical propriety of the gum may allow plant breeders to develop improved varieties of carob which can economically benefit commercial cultivation.

The significant biochemical differences observed provide a basis for the selection of plant material with desired traits from different populations.

Biochemical characterisation of fruit quality offers new opportunities to evaluate important fruit postharvest traits (Rudell, 2010). Standardized trait evaluation among breeding programs and, most importantly, germplasm collections is expected to allow more precise comparison between populations, expediting integration of economically important fruit quality traits into commercial populations (Rudell, 2010).

The principal components analysis showed interesting relationships between pulp and locust bean gum characters. The group of populations 7 and 9 from Upper and Low semi arid bioclimates, respectively, was characterized by high pulp moisture, Ash, protein, phenols, acid insoluble matter and neutral and acid detergent fibre contents. The locust bean gum in these populations was rich on moisture, protein, acid insoluble matter and phenols contents and had higher rate of mannose / galactose. They also had a good quality of gum that is characterized by a high viscosity. The second heterogenic group contains populations 2, 3, 4 and 5 from the sub humid zone and the population 6 from the Upper semi arid bioclimate. This group showed fruits and gum with moderate characteristics.

Populations 1, 10 and 11, clustered together, were characterized by a high quality of gum with high yield and viscosity. Contrary to this group, population 12 (from the Upper arid) was characterized by fruits with higher contents of saccharose, fructose, glucose and protein and a gum rich only on ash and acid detergent fibre. The quality and viscosity of gum is not good, but population 8 (from the Lower semi arid zone) had a good quality of gum with high viscosity and mannose / galactose ratio. The pulp was characterized by very high pulp moisture, ash, protein, phenols, acid insoluble matter and neutral and acid detergent fibre contents. The locust bean gum was very rich on moisture, protein, acid insoluble matter and phenols contents.

The structuring of populations according to their biochemical characteristics depends on bioclimate rather than geographic distance between sites. The UPGMA clustering established for all populations through Euclidean distances did not clearly show that, for the majority of populations, grouping had resulted from geographic location.

The biochemical variation among populations could result both from genetic and environmental factors. However, the variation observed among populations from the same bioclimatic zone, for example populations from sub humid, suggests that at least some of this variation could be related to genetic factors. The cultivation of individuals (via cuttings or grafting) in the same conditions (clonal orchard) with mixed individuals (male and female) could help the interpretation of results and facilitate selection of select economically important cultivars.

4.2 Conservation

Fragmentation of carob habitats in Tunisia adds to the spatial isolation of populations, which may reduce gene flowand increase differentiation between them (Afif et al., 2006).

The decline of populations stemming from habitat fragmentation induced genetic bottlenecks and increased genetic drift (Booy et al., 2000) may lead, particularly in disturbed sites, to gradual reduction of individual fitness (Clegg, 1995) and further population decline

Analysis of biochemical diversity and the level of differentiation between Tunisian carob populations from different geographic and bioclimatic zones is the first step in developing in-situ and ex-situ conservation strategies. It also helps provide insight into the evolutionary and demographic history of the species, and identifies potential genotypes for industrial and pharmaceutical use.

In this study, biochemical analysis has been used to detect variation between populations, and to assess their differentiation level. We suggesttargeting for conservation and for intensive cultivation and industrial use the populations with high biochemical pulp and kernels quality (populations 7 and 9) and high yield of seeds and LBG (populations 1, 10 and 11). Populations 1 and 5 should be better protected because they have a good quality of gum.

Smalland isolated populations such as populations 8 and 12 are likely the most threatened.oWe suggest that 1) these carob populations be fully protected by land use authorities, and 2) that additional populations be established through transplants to nearby suitable habitat areas.

In-situ conservation strategies should strive to preserve high biochemical diversity found in wild populations. Conservation strategies should be implemented that take into account the varied environmental conditions of each bioclimatic zone.

Our research is based not only on biochemical analysis, but also on observed high phenotypic variability within populations (size and number of seeds per pod, pod size, etc.). However, correlation between these quantitative traits or quantitative traits and molecular markers should be investigated. The joint examination of all these traits is highly recommended for the conservation of genetic resource species (Sagnard et al. 2002). Further genetic diversity analyses combining biochemical and adaptive traits are needed to assist in developing more fully conservation and management strategies for the species.

The analysis of biochemical diversity of Tunisian carob populations has led to useful information which could help preserve the genetic diversity of the species. Ex-situ preservation should be based on a maximum number of individuals collected within populations in each ecological group and their propagation in different bioclimates by means of cuttings.

5. Conclusion

The nutritive value of different fractions of carob pods and kernels was evaluated by their chemical composition. There were significant biochemical differences among the fractions of carob pods in relationship with bioclimatic location of populations.

Carob pulp was characterised by high sugar content, relatively moderate protein content compared with kernels farina. Additionally, it was established by Craig et al. (1984) that carob is free of the two anti-nutrients found in cocoa (theobromine and caffeine).

The data obtained in this study show the high variability of the carob pods and kernels collected in different areas of Tunisia. These results suggest the importance of preserving the genetic resources of carob to elaborate improvement programs, with the aim of the clonal selection with highest yield of seeds and LBG for intensive cultivation and industrial use.

Development of varieties of carob trees better suited to actual and future demands of industry must take in consideration a strategy to prevent genetic erosion in the wild. Evaluation of biodiversity in non-grafted or wild carob trees throughout the region is fundamental to successful conservation.

6. Acknowledgments

The authors thank the staff of the Institute of Research in Rural Engineering, Water and Forestry (INRGREF, Tunis) and the Wallon Centre of Industrial Biology (CWBI). University of Liege Sart Tilman for their help during the field work.

7. References

Afif, M., Benfadhl, N., Khouja, M.L. & Boussaid, M. (2006). Genetic diversity in Tunisian *Ceratonia siliqua* L. (*Cesalpinioideae*) natural populations. *Genetic Resources and Crop Evolution*, 53, 1501-1511.

Albanell, E., Caja, G. & Plaixatas, J. (1991). Characteristics of Spanish carob pods and nutritive value of carob kibbles. *Options méditerranéennes*, 16, 135-136.

Anderson, D.M.W. 1986. Food Additives and Contaminants 3: 231-234.

Andrade, C.T., Azero, E.G., Luciano, L. & Gonçalves, M. P. (1999). Solution properties of the galactomannans extracted from the seeds of *Caesalpinia pulcherrima* and *Cassia javanica*: comparison with locust bean gum. *International Journal of Biological Macromolecules*, 26, 181-185.

AOAC. (1984). Official Methods of Analysis (13th edition). Association of Official Analytical Chemists, Washington DC. USA.

AOAC. (1990). Official Methods of Analysis (15th edition). Association of Official Analytical Chemists, Washington DC. USA.

Bengoechea, C., Romero, A., Villanueva, A., Moreno, G., Guerrero, A., Puppo, M.C. (2008). Composition and structure of carob (*Ceratonia siliqua* L.) germ proteins. . *Food chem.*, 107:675- 683

Biner, B., Gubbuk, H., Karhan, M., Aksu, M. & Pekmezci, M. (2007). Sugar profiles of the pods of cultivated and wild types of carob bean (*Ceratonia siliqua* L.) in Turkey. *Food Chem, 100*, 1453-1455.

Blenford, D.E. 1979. Toasted carob powder. *Confectionery Manufacture and Marketing, 16*, 15-19.

Booy, G., Hendriks, R.J.J., Smulders, M.J.M., Van Groenendal, & D. Vosman B. (2000). Gene diversity and the survival of populations. *Plant Biol*. 2: 379–395.

Bouzouita, N., Khaldi, A., Zgoulli, S., Chebil, L., Chaabouni, M. & Thonart, P. (2007). The analysis of crude and purified locust bean gum: A comparison of samples from different carob tree populations in Tunisia. *Food chem.*, 101, 1508-1515.

Brandt, E. (1984). Carob. *Nutr and food Sc, 91*, 22-24.

Carlson, W.A. (1986). The carob: evaluation of trees, pods and kernels. Intl. *Tree Crops J*, 3,281-290.

Clegg, M.T. (1995). Conserving biological diversity: an evolutionary perspective. *Calif. Agric.* 49: 34–39.

Craig, W.J. & Nguyen, T.T. (1984). Caffeine and theobromine levels in cocoa and carob products. *Journal of Food Sc*, 49, 302-306

Doublier, J. L. (1975). Propriétés rhéologiques et caractéristiques macromoléculaires des solutions aqueuses de galactomannanes. *PhD Thesis*. Université Paris VI France.

Egli, W. (1969). Origin, characteristics, and possible applications of plant hydrocolloids, particularly galactomannan. *Schweizerusche Zeitschriftfuer Obst-und Weinbau, 105,* 470- 477.

Emberger, L. (1966). *Une classification biogéographique des climats.* Recherches et Travaux de Laboratoires de Géologie, Botanique et Zoologie. Faculté des Sciences de Montpelier (France), 7, 1-43.

Food and Nutrition Board-National Research Council. (1981). *Food chemicals codex.* Washington, DC: National Academy Press.

Fulgancio, S., Calixto, F.S. & Canellas, J. (1982). Components of nutritional interest in carob pods *Ceratonia siliqua. Journal of the Science of Food Agr, 33,* 1319-1323.

Imeson, A. (1997). *Thickening and gelling agents for food* (2nd ed.). New York: Culinary and Hospitality Industry Publication Services.

Kapoor, V.P. (1972). *Phytochem, 11,* 1129- 1132.

Lee, C.Y., Kagan, V., Jawarski, A.W. & Brown, S.K. (1990). Enzymatic browning in relation to phenolic compounds and polyphenol oxidase activity among various peach cultivars. *Journal of Agricultural and Food Chem, 88,* 99-101.

Lopes da Silva, J. A. & Gonçalves, M. P. (1990). Food Hydrocolloid, 4, 277–287.

McCleary, B. V. & Matheson, N. K. (1976). Galactomannan utilization in germinating legume seeds. *Phytochem, 15,* 43–47.

Mhaisen, A. (1991). Cocoa tree. *Agric. Eng.* 43, 90–91.

Rafique, C.M. & Smith, F. (1950). Processing and characterization of carob powder. *Food Chemistry, 69,* 283-287.

Rudell, D. (2010). Standardizing postharvest quality and biochemical phenotyping for precise population comparison. *Hortsc,* 9, 1307-1309

Sagnard, F., Barberot, C. & Fady, B. (2002). Structure of genetic diversity in Abies alba Mill. From southwestern Alps: multivariate analysis of adaptive and non-adaptive traits for conservation in France. *Forest Ecol. Manage,* 157: 175–189.

Silanikove, N., Landau, S., Or, D., Kababya, D. & Nitsan, Z. (2006). Analytical approach and effects of condensed tannins in carob pods (*Ceratonia siliqua*) on feed intake, digestive and metabolic responses of kids. *Livestock Sc, 99,* 29-38

Tamir, M. & Nachtomi, E. (1973). Urinary phenolic metabolites of rates fed carobs (*ceratonia siliqua*) and carob fractions. *Int J Biochem, 3,* 123-124.

Tous, J., Rovira, M., Romer, A., Afif, M., Khouja, M.L., Naghmouchi, S. & Boussaid, M. (2006). Carob tree germplasm in Tunisia. FAO-CIHEAM. *Nucis newsletter,13,* 55-59.

Tous, J., Rovira, M., Romer, A., Afif, M., Khouja, M.L., Naghmouchi, S. & Boussaid, M. (2006). *Estudio de poblaciones de algarrobo en Tunez.* III congreso de mejora genetica de plantas Valencia. *Actas de horticultura, 45,* 233-236.

Yousif, A.K. & Alghzawi, H.M. (2000). Processing and characterization of carob powder. *Food Chem, 69,* 283–287.

2

Protected Areas in Selected Arab Countries of the Levant Region (Syria, Lebanon & Jordan): An Evaluation of Management and Recommendations for Improvement

Brandon P. Anthony and Diane A. Matar
Environmental Sciences & Policy Department,
Central European University, Budapest,
Hungary

1. Introduction

Global trends in biodiversity conservation have frequently been reported as being unsatisfactory, especially after the 2010 targets of the Convention on Biological Diversity (CBD) failed to be met (2010 Biodiversity Indicators Partnership 2010). Despite some notable conservation successes at various scales (Sodhi et al., 2011), anthropogenic impacts go largely unabated and increasingly endanger the planet's biota and life support systems (Dirzo & Raven, 2003). One of the main approaches to halting biodiversity loss has been the establishment of protected areas (PAs), an undertaking which has seen a prolific growth in recent decades in terms of both number and spatial extent (Chape et al., 2005; Coad et al., 2008a). While the number of PAs under national or international programs and legislation has been rising on a global level (Butchart et al., 2010; Coad et al., 2008b), biodiversity loss continues even within some PAs (Bonham et al., 2008; Craigie et al., 2010; Gaston et al., 2008; Hockings & Phillips, 1999; Oates, 1999). Why is this?

While the answer to this question is complex, one important factor being closely investigated is the effectiveness level of PAs management (Cantu-Salazar & Gaston, 2010; Mulongoy & Chape, 2004). It is now clear that the effectiveness of PAs in conserving biodiversity cannot be inferred simply as a result of their number and size, but also depends on their location, structure (shape, connectivity to other sites, etc.) and, of equal importance, their management (Anthony & Szabo, 2011; Rodrigues et al., 2004). Many evaluation tools have been developed for assessing and monitoring PA management effectiveness, many of which are based on the International Union for Conservation of Nature - World Commission on Protected Areas (IUCN-WCPA) Framework and are now commonly used worldwide (Ervin, 2003; Leverington et al., 2008; WWF, 2007).

Three Arab countries of the Levant region: Syrian Arab Republic (Syria), Hashemite Kingdom of Jordan (Jordan), and Lebanese Republic (Lebanon), are part of the Mediterranean Basin hotspot area for conservation (Mittermeier et al., 2004; Myers et al.,

2000). Given the high global conservation value of their fauna and flora, and their complex socio-political and economic contexts, these countries offer an excellent opportunity for biodiversity research. The rich historical background and turbulent political situation of the area has sometimes negatively influenced the degree of national or international attention given to nature protection. However, in recent decades, more sustained efforts have been made to create well-defined, legally recognized PAs in the region. While the three countries are geographically related, they present many differences in their ecosystems, national governance, and PAs establishment and management systems.

The call by Hockings et al. (2006: viii) to "look for common threads... to find trends, themes and lessons across regions" is particularly relevant in our study, as there is a paucity of documented data on PA management effectiveness evaluation in this region. Our research provides a valuable 'snapshot' evaluation of the current status of management of established PAs and UNESCO Biosphere Reserves in Lebanon, Jordan and Syria based on data collected in September 2011, during the 'Arab Spring' period, with Syria being most seriously impacted at this time. Our evaluation method is based on the thirty-three indicators developed by Leverington et al. (2010) that provide a practical and comprehensive approach for a quick evaluation of PA management effectiveness. This chapter provides a critical review of the current situation in the Levant region and compares it with the global results reported by Leverington et al. (2010). Here, we address three pertinent questions:

1.   How effective is protected area management?
2.   Which aspects of management are most effective?
3.   Which factors are most related to (a) overall effectiveness, and (b) successful outcomes?

The results of this comparison are then used to devise recommendations for improving the management of PAs in the Levant region, which we hope will contribute to improving the conservation of its unique biodiversity.

2. Management effectiveness of protected areas in global agendas

Management effectiveness evaluation (MEE) is defined by Hockings et al. (2006: xiii) as "the assessment of how well the PA is being managed – primarily the extent to which it is protecting values and achieving goals and objectives. The term management effectiveness reflects three main themes:

- design issues relating to both individual sites and PA systems;
- adequacy and appropriateness of management systems and processes; and
- delivery of PA objectives including conservation of values."

The absence of a coherent, unified set of indicators to measure PA effectiveness in reaching conservation goals, combined with the significant rise in global impacts of human activities on PA conservation capacity, created an 'urgent' need to improve PA management effectiveness within the short (2010) deadline of the CBD agenda (Chape et al., 2005). As reported by IUCN, "Many protected areas around the world are not effectively managed. In response, management effectiveness will continue as a priority with a focus on improving on and learning from past approaches" (IUCN-WCPA, 2009: 1).

Protected Areas in Selected Arab Countries of the Levant Region (Syria, Lebanon & Jordan): An Evaluation of
Management and Recommendations for Improvement

19

Many initiatives were undertaken towards this aim, for example, as part of the CBD's 7th Conference of Parties (COP-7) Programme of Work on Protected Areas (PoWPA) in 2004, nations committed to develop assessment systems to report on PA effectiveness for 30% of their PAs by 2010 (WWF, 2007), a commitment that was subsequently increased to 60% by 2015 (CBD 2010). A second initiative was adopted at the CBD/COP-8 meeting in 2006, where delegates reviewing the first PoWPA implementation phase highlighted the need to improve PA management effectiveness by tackling the following underlying issues: (i) lack of financial resources; (ii) lack of technical assistance and capacity-building for PA management staff; (iii) poor governance; and (iv) political, legislative and institutional barriers (SCBD, 2009; UNEP, 2006). In response, the purposes underlying the development of management effectiveness evaluation were that it should lead to improved management in changing environments, more effectively allocate resources, enhance transparency and accountability, and build constituency by involving the community and promoting PA values (Hockings et al., 2006).

Further, as part of technical assistance and capacity building, one solution highlighted by international experts was to create cost-effective evaluation tools for monitoring progress towards management targets. As underscored in the Durban Congress recommendations: "New methodologies to assess management effectiveness should be developed to address the specific gaps identified […] including rapid, site level assessments of both management effectiveness and threats" (IUCN, 2005: 92). Actions taken in this perspective include the development by the IUCN-WCPA of a 'Protected Areas Programme' which partially aimed at providing capacity-building for increasing management effectiveness of PAs through the provision of guidance, tools and other information, and a vehicle for networking (IUCN-WCPA, 2009).

3. Monitoring tools

Monitoring has been best described as the collection and analysis of repeated observations or measurements to evaluate changes in condition and progress toward meeting a management objective (Elzinga et al., 2001; Tucker, 2005). As one essential component of adaptive management (Holling, 1978; Salafsky et al., 2001; Tucker, 2005), monitoring involves a continuous evaluation of progress towards project goals including the preservation of species from internal or external threats (Margules & Pressey, 2000). Monitoring is also an essential part of systematic conservation planning as it constitutes the last of six stages as defined by Margules & Pressey (2000).

Several tools and indicators have been developed by international organizations and experts to evaluate PA management effectiveness (Leverington et al., 2008). Some of the most widely used include the Management Effectiveness Tracking Tool (METT) (WWF, 2007), Rapid Assessment and Prioritization of Protected Area Management (RAPPAM) (Ervin, 2003), and Threat Reduction Assessment (TRA) (Salafsky & Margoluis, 1999; Anthony, 2008). However, as different PA sites and networks have diverse characteristics (e.g. management structure, geographical coverage and variation) and are embedded within various cultural, political and socio-economic contexts, there is no one standard tool that is globally accepted so far (Chape et al., 2005). Consequently, the tool chosen for monitoring management effectiveness should be adapted to the specific settings, capacities, needs and objectives of the PA or PA network in which it will be applied.

3.1 World Commission on Protected Areas (WCPA) framework

The IUCN-WCPA task force responded to the need for management effectiveness tracking tools by developing a framework in 1997 that aims at providing overall guidance in the development of more adapted assessment systems and to encourage the presence of standards for assessment and reporting (Hockings et al., 2000; WWF & WB, 2003). The WCPA Framework was developed on the concept that good PA management is based on six elements: context, planning, inputs, processes, outputs, and outcomes (see Table 1).

Element of Evaluation	Explanation	Criteria assessed	Focus
Context	*Where are we now?* Evaluation of importance, threats & policy environment	Significance Threats Vulnerability National context Partners	Status
Planning	*Where do we want to be?* Evaluation of PA design & planning	PA legislation & policy PA system design Management planning	Appropriateness
Inputs	*What do we need?* Evaluation of resources needed to carry out management	Resourcing of agency Resourcing of site	Resources
Processes	*How do we go about it?* Evaluation of way in which management is conducted	Suitability of management actions	Efficiency & appropriateness
Outputs	*What were the results?* Evaluation of implementation of management programs & actions Delivery of products & services	Results of management actions Services & products	Effectiveness
Outcomes	*What did we achieve?* Evaluation of outcomes & the extent to which they achieved objectives	Impacts/effects of management in relation to objectives	Effectiveness & appropriateness

Table 1. Summary of the IUCN-WCPA Framework (adapted from Hockings et al., 2006).

In summary, the cycle starts by an understanding of the context of values and threats present in the PA. It then progresses through planning, allocating resources and processing management actions. These result in products and services that have a final impact on management objectives (Hockings et al., 2006; WWF, 2007; WWF & WB, 2003). The WCPA Framework also stresses the importance of establishing clear, measurable, and outcome-based objectives as a basis for the whole management process and for better monitoring of results (MacKinnon et al., 1986; Tucker, 2005). The WCPA provided the first consistent scheme to monitoring PA management effectiveness, and has been used by many other experts/organizations to develop specific assessment tools (e.g. METT and RAPPAM).

Based on the plethora of scoring and monitoring methodologies, Leverington et al. (2010) compiled over 8000 assessments from more than 50 methodologies to develop a common

scale and list of 33 'headline indicators'. These indicators are categorized according to the six
evaluative elements embedded within the IUCN-WCPA Framework (see Table 1), and serve
as the indicators utilized in our own study.

4. Conservation values and protected areas in Syria, Lebanon, and Jordan

4.1 The Levant region

The word Levant comes from the French language meaning 'rising'. After World War I, the
French Mandates of Syria and Lebanon (1920-1946) were called the Levant States but the word
now mostly refers to the geographic and cultural zone of West Asia bounded by the Syrian
Desert to the east, Mediterranean Sea to the west, Taurus Mountains to the north, and
the Arabian Desert to the south. Nowadays, the Levant refers to most of modern Syria,
Lebanon, Jordan, Palestinian Territories, Israel, and sometimes parts of Turkey and Iraq. It is a
more or less heterogeneous region, divided into areas of diverse ecological and environmental
character close to that of southern California (Living University, 2009; Sabatinelli, 2008).

Syria, Lebanon and Jordan are three neighboring countries of the East Mediterranean Basin,
which differ in their number and extent of formal reserves and biosphere reserves (Table 2).
Syria and Lebanon are bordered by the East-Mediterranean coast on their west side, while
Jordan is further situated inland and separated by Israel and Palestinian lands to the
Mediterranean Sea (Fig.1).

Country	Area (km²)	Population[a]	No. of PAs[b]	% coverage of PAs
Jordan	89,342	6,508,271	9[c]	1.7
Lebanon	10,451	4,143,101	13[d]	6.2
Syria	185,180	22,517,750	27[e]	1.4

[a] most recent estimate, according to www.cia.gov/library/publications/the-world-factbook/geos/
[b] for definition of PA used in our study, please see section 5.1.1.
[c] http://www.rscn.org.jo and http://www.aqabazone.com/
[d] MOE-L et al. 2011
[e] SAR et al. 2009

Table 2. Characteristics of countries included in this study.

4.2 Conservation values of the region

Syria, Lebanon and Jordan are countries with high conservation values within the
Mediterranean Basin hotspot area. The Mediterranean Basin, stretching from northern Italy
to Morocco, and from Portugal to Jordan, has been recognized as an international hotspot
area for biodiversity (CI, 2007; Myers et al., 2000). This hotspot region hosts about 22,500
endemic vascular plant species, more than four times the total amount found in the rest of
Europe (CI, 2007).

A global hotspot analysis of the 5 regions in the world with a Mediterranean climate
identified 10 red alert hotspot areas in the Mediterranean Basin, one of which includes
Lebanon and Syria (Medail & Quezel, 1997, 1999). This area is characterized by a high level
of plant richness and endemism (Medail & Quezel, 1999; Talhouk & Abboud, 2009). The
historical high level of anthropogenic threats in the Mediterranean region has been
pressuring the natural diversity and threatening its persistence, making it a hotspot area

Fig. 1. Map showing Syria, Lebanon and Jordan as part of the Levant and East-Mediterranean region.

under threat (CI, 2007; Cuttelod et al., 2008). The IUCN's Redlist classifies 143 species as "Threatened" in the 3 countries in total, of which 82 are vertebrates (IUCN, 2011).

The 2009 Report of the Arab Forum on Environment and Development, covering 20 Arab countries, reported Lebanon and Syria as two of the countries with the richest biodiversity in the Arab world with recorded numbers above 3000 and 5000 (species/country) for flora and fauna, respectively (Talhouk & Abboud 2009).

Lebanon has one of the highest densities of floral diversity in the Mediterranean Basin, which is in turn considered one of the most diverse regions in the world. Lebanese biodiversity includes 4633 flora and 4486 fauna species of which many are threatened (MOE-L et al., 2009). Syria lists 3300 flora species and more than 3300 fauna species on land and in water (SAR et al., 2009). Jordan hosts more than 2500 species of flora and while the total number of fauna species is not reported, more than 75 species of mammals, 425 birds, 450 fish, and 102 reptiles and amphibians have been mentioned in the Fourth National Report to the CBD (MOE-J, 2009).

Moreover, Lebanon has a remarkably high flora species/area ratio of 0.25 species/km^2 compared with 0.022 for Jordan, and 0.017 for Syria (MOE-L et al., 2009). The faunal diversity of Lebanon is also relatively higher than Syria and Jordan with a ratio of 0.028 species/ km^2

compared with 0.019 and 0.015 for Syria and Jordan, respectively (MOE-J, 2009; MOE-L et al., 2009). Despite their international conservation value, Syria, Lebanon and Jordan have only relatively recently focused their efforts on improving biodiversity conservation through the creation of PAs. These neighboring countries present many differences in their PAs management and monitoring systems as they are at different stages of PA evolution. Given the economic and political context of these countries and the lack of research on PAs, they represent interesting case-studies in the Arab and international arena. From a national and political perspective, they share a regional atmosphere of political instability, and a common lack of national prioritization for biodiversity conservation.

4.3 Protected areas

4.3.1 Jordan

Jordan currently includes nine formally recognized reserves (Table 3), with two designated as Biosphere Reserves (Dana, Al-Mujib) (UNESCO, 2011). The management of these sites has developed under several conservation projects; however the Fourth National Report to the CBD in Jordan still reports many obstacles to effective conservation encompassing PAs, including 'Incomplete national guidelines and management plans for conservation sites', and the 'lack of a national knowledge management and data processing system for monitoring and reporting on biodiversity' (MOE-J, 2009: 15). Seven PAs in Jordan are managed by the Royal Society for the Conservation of Nature (RSCN) in agreement with the Ministry of Environment (MOE). The other two PAs (Aqaba Marine Park, and Wadi Rum Protected Area) fall under the direct management of Aqaba Special Economic Zone (ASEZA) (RSCN, 2008).

A report by RSCN (2008) presents the results of an evaluation carried out on all 8 PAs in Jordan (at that time) to assess their management effectiveness for the first time since their establishment. The evaluation was done through a joint effort between the managing staff of the reserves, RSCN, ASEZA and IUCN local office experts, using the METT tool. Results reflected an "acceptable level of management effectiveness for all sites" (RSCN, 2008: 4) however, in some cases, there was a clear difference in the management effectiveness scores between sites. All six elements of the METT tool: context, planning, inputs, process, outputs and outcomes, were analyzed relatively to the overall score, consistently showing positive influence on the final score (RSCN, 2008).

Recommendations for improvement were consistent with the Fourth National Report to the CBD, demanding greater official recognition and integration of the PA network and related resource management policies into national strategies and action plans (MOE-J, 2009). Moreover, more effective national bylaw drafting and finalization was requested for issues relating to PA threats such as hunting. The strengthening and systemization of management plans' monitoring and evaluation was also recommended in order to provide more rapid feedback to PAs management teams and to allow more effective adaptive management practices (RSCN, 2008).

4.3.2 Lebanon

The official and legal designation of PAs in Lebanon began in 1992 when the first two Nature Reserves were designated: Horsh Ehden (mixed forest), and Palm Islands (marine reserve). The *State and Trends of the Lebanese Environment 2010* report recognizes 10 legally established Nature Reserves (marine, coastal and mountain ecosystems) under the jurisdiction of the MOE in Lebanon, which cover approx. 2.2% of the Lebanese territory (MOE-L et al., 2011). In addition, there are three internationally recognized Biosphere Reserves (Shouf, Jabal Rihane, Jabal Moussa), of which Shouf is also (partially) a Nature Reserve. The management of these PAs in Lebanon relies mainly on managing institutions' projects funds, although for Nature Reserves, funding is also allocated annually from the MOE. Despite several PAs having developed management and/or monitoring plans, the effective implementation of these plans is often hindered by the lack of technical skills and resources, or minimal follow-up by the national managing teams. Moreover, the absence of a national monitoring plan remains a major impediment for effective biodiversity conservation (Matar & Anthony, 2010).

Concerning monitoring and evaluation of conservation efforts in Nature Reserves, plans in Lebanon have been focused so far on the use of biological indicator species and Geographic Information Systems (MOE-L, 2002; MOE-L & LU 2004a, 2004b; UNDP, 1995, 2005), which has led to an improvement in reporting of species and habitats, and area coverage. However the monitoring pace has been slow and unsustainable due mostly to limited funds and project dependency (UNDP, 2005). The need to have a cost-effective tool to monitor management effectiveness was identified and was partially addressed by the MOE under the Stable Institutional Structure for Protected Areas Management (SISPAM) project which developed an adapted version of METT for Lebanese Nature Reserves management monitoring (Hagen & Gerard, 2004; MOE-L, 2005, 2006a, 2006b). Yet, political turmoil and the resultant governmental instability after the 2006 war has retarded the ratification of the decision to implement the SISPAM outcomes (including the adapted METT monitoring tool), leaving the choice and implementation of METT (or similar tools) up to individual PAs.

4.3.3 Syria

According to the latest Syrian report for the CBD, there are 27 legally established PAs in Syria covering 1.4% of the country's territories, including the Lajat Biosphere Reserve established in 2009 (SAR et al., 2009). Most PAs still lack an effective management system and a biodiversity monitoring strategy (SAR et al., 2009). Since 2004, and in the scope of a UNDP-GEF 'Biodiversity Conservation and Management Project', only three PAs have been developed: the Abou-Qubies in central-northwestern Syria, the Al Fourounloq (or Furunloq) in the northwestern coastal region of Syria and Jebel Abdul Aziz in northeastern Syria (UNDP, 2004; SAR et al., 2009). Through this project, management practices focusing on the participation of local communities were emphasized and established for the three reserves (UNDP, 2004). Further, the 2009 CBD report highlighted the imminent need for a more thorough identification of biodiversity hotspots within the Syrian borders with the aim to extend the PA system and improve coverage of important ecological sites (SAR et al., 2009). On the other hand, the absence of effective management programs and of a national monitoring strategy was highlighted as a priority for the Syrian Arab Republic's government. Management of all PAs (including the Lajat BR) in Syria remains a centralized process under the Ministry of State for Environmental Affairs (SAR et al., 2009).

5. Methods

5.1 Data collection

5.1.1 Survey and response levels

In addition to archival research, which was based on published data concerning PAs in Syria, Lebanon, and Jordan (including Fourth National Reports to CBD), we prepared and conducted an evaluation survey sent by email to identified representatives of PAs' managing institutions (direct management teams), who were judged to be the most appropriate respondents to complete the evaluation given their familiarity with the site and direct on-site management experience. Criteria for PA selection was based on the adopted definition of a PA for this study, i.e. "a formal reserve or biosphere reserve, recognized nationally and/or internationally".

Accordingly, 9 PA representatives were contacted in Lebanon, of which 8 responded (Table 3). The respondents' sample includes six Nature Reserves designated by law from the MOE, one of which is included in a Biosphere Reserve (Shouf), and two other Biosphere Reserves. In Jordan, representatives from 7 RSCN PAs responded, including two Biosphere Reserves. For Syria,

Name of Protected Area	Year of National Designation	International Designation (year)	Area (ha)
Jordan			
Ajloun Forest Reserve	1989		1200
Azraq Wetland Reserve	1978		1200
Dana Biosphere Reserve	1993	BR (1998)	29,200
Dibeen Forest Reserve	2005		850
Al-Mujib Nature Reserve	1987	BR (2011)	21,200
Shaumari Wildlife Reserve	1987		2200
Yarmouk Nature Reserve	2010		206
Lebanon			
Al Shouf Cedar Nature Reserve / Shouf Biosphere Reserve	1996	BR + IBA (2005)	16,000 (NR) 50,000 (BR)
Bentael Nature Reserve	1999	IBA	110
Horsh Ehden Nature Reserve	1992	IBA	1100
Jabal Moussa	2008	BR + IBA (2009)	6500
Jabal Rihane	2006	BR (2007)	11,300
Palm Islands Nature Reserve	1992	IBA, Ramsar site, SPA	415
Tannourine Cedar Forest Nature Reserve	1999	IBA (2006)	620
Tyre Coast Nature Reserve	1998	Ramsar site (1999)	380
Syria			
Abou-Qubies	1999		4500
Al Fourounloq	1999		5390
Jebel Abdul Aziz	1993		49,000

Note: BR=Biosphere Reserve, IBA=Important Bird Area, SPA=Specially Protected Area

Table 3. Protected areas included in the study, their year of designation, and area.

data collection was very difficult since the country was engaged in a political crisis during the time of the survey, with a major revolution against the regime; hence the response rate was very low (3 of 27 PAs).

5.1.2 Survey questionnaire

The survey questionnaire was based on the 33 indicators developed by Leverington et al. (2010) which comprehensively summarize reviewed indicators from all Protected Areas Management Effectiveness (PAME) methodologies (Table 4). The indicators are grouped into

Element	Headline Indicator
Context	Level of significance
	Extent and severity of threats
	Constraint or support by external political and civil environment
Planning	Protected area gazettal (legal establishment)
	Tenure issues
	Adequacy of protected area legislation and other legal controls
	Marking and security or fencing of park boundaries
	Appropriateness of design
	Management plan
Input	Adequacy of staff numbers
	Adequacy of current funding
	Security/reliability of funding
	Adequacy of infrastructure, equipment and facilities
	Adequacy of relevant and available information for management
Process	Effectiveness of governance and leadership
	Effectiveness of administration including financial management
	Management effectiveness evaluation undertaken
	Adequacy of building and maintenance systems
	Adequacy of staff training
	Staff/other management partners skill level
	Adequacy of human resource policies and procedures
	Adequacy of law enforcement capacity
	Involvement of communities and stakeholders
	Communication program
	Appropriate program of community benefit/assistance
	Visitor management (visitors catered for and impacts managed appropriately)
	Natural resource and cultural protection activities undertaken
	Research and monitoring of natural/cultural management
	Threat monitoring
Outputs	Achievement of set work program
	Results and outputs produced
Outcomes	Conservation of nominated values – condition
	Effect of park management on local community

Table 4. The 33 indicators used in the common PAME reporting format, according to evaluation element.

the 6 evaluation elements of the WCPA Framework, which also constitute the most effective framework for quantitative evaluations of PA management effectiveness (Hockings, 2003).

In contrast to Leverington et al.'s (2010) scoring on a scale from 0 to 1, respondents in our study were asked to allocate a score to each and all indicators on a scale from 0 to 10, where 0 represented the lowest measurement (0 = no management at all/no progress) and 10 represented the optimum situation (10 = high management standards/ideal situation achieved).

Given the subjective nature of *scoring* (in contrast to *monitoring*), scores are allocated qualitatively, are perception-based, and therefore are only estimates of progress (Cook & Hockings, 2011; Hockings, 2003). Thus, we recognize this limitation and interpret our results with caution, especially in the absence of complementary quantitative data. Nevertheless, the utility of this scoring does allow a rapid 'snapshot' self-evaluation of PA management status based on which recommendations for improvement can be derived.

5.2 Data analysis

Quantitative data were analyzed using IBM® SPSS® Statistics (ver. 19). Both univariate and bivariate descriptive statistics were used, including measures of central tendency and dispersion, and Pearson's Correlation when exploring correlations between interval level variables. When comparing means, z-tests were used to compare sample and population means, t-tests were utilized for two independent samples and ANOVA for three or more samples. If ANOVA indicated significant mean differences, Scheffe post hoc tests which are appropriate when sample sizes are unequal, were used to identify which means differed (Scheffé, 1953). Alpha level for all tests was set at 0.05. We present national data as aggregates and compare countries to one another, and also to the global results from Leverington et al. (2010), which serves as a rough benchmark for comparison.

6. Results

Overall, the management effectiveness scores across the 18 PAs in the studied region ranged from 3.58 to 9.18, with a mean score of 7.01± 1.54 (Fig. 2). This value is significantly greater ($z=4.27$, $p<.001$) than the mean of 5.30± 1.7 (adjusted based on scale difference) reported by Leverington et al. (2010), based on their global set of 3184 assessments.

From the completed questionnaires, the PA management effectiveness mean score for Jordan was 8.50± 0.72 (n=7), 6.55± 0.08 for Syria (n=3), and 5.87± 1.22 for Lebanon (n=8) (Fig. 3). Jordan's mean score is significantly greater than that of both Syria's ($t=7.045$, $p<.001$, $df=6.631$) and Lebanon's ($t=4.981$, $p<.001$, $df=13$). Moreover, only Jordan had a significantly higher mean score than the global average ($z=4.98$, $p<.001$).

When we compared the three countries according to the six evaluative elements, Jordan had significantly higher mean scores than Lebanon across all categories, except *outcomes* (see Fig. 4; Table 5). Moreover, Jordan's mean scores were significantly higher than Syria's in the *context* category. It is also noteworthy that mean scores for *output* indicators (achievement of set work program; results and outputs produced) had relatively high variability within both Jordan and Lebanon PAs.

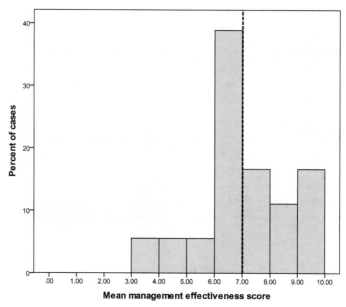

Fig. 2. Distribution of mean scores for protected area management effectiveness assessments in Jordan, Lebanon, and Syria. (Mean score across all assessments is shown as a *vertical line*; N=18).

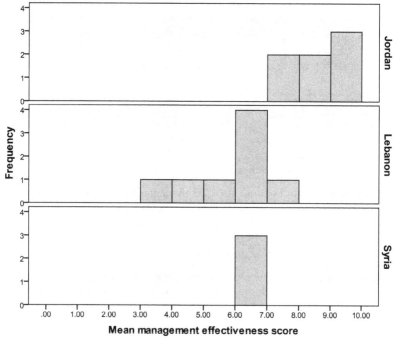

Fig. 3. Mean management effectiveness scores across the region (N=18).

Dependent Variable	(I) Country where PA located	(J) Country where PA located	Mean Difference (I-J)	Std. Error	Sig.
Mean score of *context* indicators	Jordan	Lebanon	1.929	.4085	.001
		Syria	2.651	.5446	.001
Mean score of *planning* indicators	Jordan	Lebanon	2.458	.5355	.001
Mean score of *input* indicators	Jordan	Lebanon	2.639	.6602	.004
Mean score of *process* indicators	Jordan	Lebanon	2.913	.6029	.001
Mean score of *outputs* indicators	Jordan	Lebanon	2.982	.9376	.021

Table 5. Multiple comparisons of mean scores of evaluative elements across the three
countries. Only significant mean differences are shown, based on Scheffe post hoc tests.

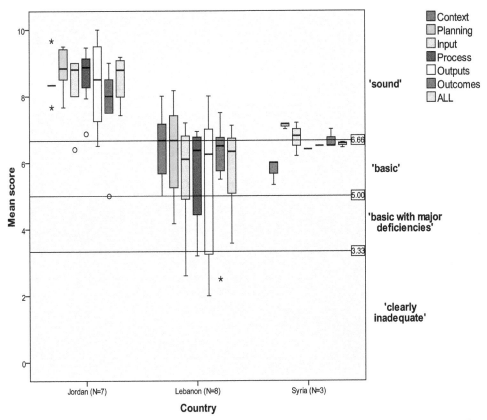

Fig. 4. Distribution of mean scores for each country across the six evaluative elements, and
overall mean score. Note: Mean scores <3.33='clearly inadequate management'; 3.33-
5.00='basic management with major deficiencies'; 5.01-6.66='basic management';
>6.66='sound management'.

When we calculated the mean scores for the 33 headline indicators, clear patterns emerged
in which 4 of 6 *planning* indicators were among the 7 highest scoring indicators (Table 6).

Headline Indicator	Element	Mean	S.D.
'Sound management' [score >6.66]			
Protected area gazettal (legal establishment)	Planning	8.67	2.114
Level of significance	Context	8.39	.916
Adequacy of relevant and available information for management	Input	8.11	1.079
Appropriateness of design	Planning	8.06	.873
Tenure issues	Planning	7.94	2.127
Research and monitoring of natural/cultural management	Process	7.83	1.790
Adequacy of protected area legislation and other legal controls	Planning	7.33	1.495
Threat monitoring	Process	7.33	2.249
Conservation of nominated values — condition	Outcomes	7.17	1.618
Management effectiveness evaluation undertaken	Process	7.11	2.423
Involvement of communities and stakeholders	Process	7.06	2.209
Effectiveness of governance and leadership	Process	7.06	1.697
Management plan	Planning	7.06	2.508
Staff/other management partners skill level	Process	6.94	1.862
Natural resource and cultural protection activities undertaken	Process	6.94	2.508
Adequacy of staff training	Process	6.89	1.937
Effectiveness of administration including financial management	Process	6.83	1.654
Extent and severity of threats	Context	6.83	1.689
Results and outputs produced	Outputs	6.78	2.238
Adequacy of building and maintenance systems	Process	6.78	1.734
Communication program	Process	6.78	1.833
Adequacy of human resource policies and procedures	Process	6.67	1.749
Achievement of set work program	Outputs	6.67	2.223
Appropriate program of community benefit/assistance	Process	6.67	2.058
Adequacy of infrastructure, equipment and facilities	Input	6.67	1.680
'Basic' management [score = 5.01 – 6.66]			
Security/reliability of funding	Input	6.56	2.455
Adequacy of current funding	Input	6.56	2.229
Visitor management (visitors catered for and impacts managed appropriately)	Process	6.56	2.455
Adequacy of staff numbers	Input	6.50	2.455
Effect of park management on local community	Outcomes	6.39	2.279
Adequacy of law enforcement capacity	Process	6.28	2.347
Constraint or support by external political and civil environment	Context	6.17	2.093
Marking and security or fencing of park boundaries	Planning	5.67	3.068

Table 6. The evaluative element, mean and standard deviation (S.D.) for each headline indicator analyzed.

The only planning indicator which was clearly deficient was 'Marking and security or fencing of park boundaries', ranking last in the list and particularly problematic amongst PAs in Lebanon and Syria. Other relatively weakly scoring indicators related to funding, staffing, and law enforcement capacity, as well as 'external political and civil environment' support.

When we explored correlations between each headline indicator and the overall management effectiveness score, 7 indicators had Pearson's R values >0.90 (Table 7). In addition, we investigated which indicators were most positively correlated with the two *outcome* indicators, which reflect whether the long-term objectives are met. Two of the indicators highly correlated with the overall mean score were also highly correlated with 'conservation of values', i.e. 'Adequacy of human resource policies and procedures' and 'Appropriate program of community benefit/assistance' (see Table 7). However, indicators correlated with 'effect on community' were less clear with all R values less than 0.45, indicating relatively weak overall associations with this *outcome*. Finally, we tested whether area of PA and year of national designation were correlated with mean scores. Only the latter was found to be significantly correlated, i.e. higher effectiveness scores were found to be associated with older PAs ($R=.40$, $p=.05$).

Headline Indicator	Corr. with Mean	Corr. with conservation of values (outcome)	Corr. with effect on community (outcome)
Adequacy of human resource policies and procedures	.966	.915	.300
Appropriate program of community benefit/assistance	.941	.919	.205
Adequacy of infrastructure, equipment and facilities	.938	.844	.312
Conservation of nominated values – condition	.930	--	.173
Effectiveness of administration including financial management	.921	.846	.221
Research and monitoring of natural/cultural management	.915	.863	.291
Adequacy of protected area legislation and other legal controls	.911	.802	.357
Adequacy of staff training	.897	.851	.197
Achievement of set work program	.889	.883	.155
Staff/other management partners skill level	.888	.882	.324
Results and outputs produced	.888	.888	.145
Marking and security or fencing of park boundaries	.887	.770	.289
Communication program	.884	.886	.205
Adequacy of building and maintenance systems	.866	.790	.306
Tenure issues	.829	.738	.393
Adequacy of law enforcement capacity	.826	.808	.144

Headline Indicator	Corr. with Mean	Corr. with conservation of values (outcome)	Corr. with effect on community (outcome)
Management effectiveness evaluation undertaken	.822	.655	.237
Visitor management (visitors catered for and impacts managed appropriately)	.818	.775	.117
Threat monitoring	.805	.695	.145
Involvement of communities and stakeholders (planning, decision-making etc.)	.778	.738	.194
Security/reliability of funding	.764	.805	**.338**
Protected area gazettal (legal establishment)	.758	.722	**.444**
Adequacy of current funding	.750	.772	**.441**
Natural resource and cultural protection activities undertaken	.740	.611	.138
Adequacy of staff numbers	.739	.777	.331
Effectiveness of governance and leadership	.709	.639	.009
Constraint or support by external political and civil environment	.701	.634	.245
Adequacy of relevant and available information for management	.582	.494	-.066

Note: Indicators highly correlated ($R>0.90$) with overall mean are **bold**; the five most highly correlated items with the two *outcome* indicators are also **bold**. Only 28 highest correlations ($R>0.50$) are shown.

Table 7. Correlation of headline indicators with overall mean, and *outcomes* (conservation of values; effect on community).

7. Discussion

The discussion of our results follows the framework of the analysis made by Leverington et al. (2010) in the aim of establishing a comparison of management effectiveness results between the studied Arab Levant countries and global results.

7.1 How effective is protected area management?

Our results show that, of the three countries studied, only Jordan is performing significantly better than the global average in managing PAs. None of the analyzed PAs scored in the 'clearly inadequate' management range (<3.33), 11.1% scored in the 'basic with major deficiencies' range (3.33-5.00), 38.9% in the 'basic' range (5.01-6.66) and 50% in the 'sound' management range (>6.66) (Fig. 2); this compares with global score proportions of 13%, 28%, 37%, and 22%, respectively (Leverington et al. 2010). Jordan consistently showed management effectiveness scores in the sound management range, while Syrian scores were concentrated in the basic performance, and Lebanon showed the greatest variability encompassing all ranges. These performance levels can be interpreted in light of existing literature and context. The Jordanian PAs management seems to have a more rigorous monitoring system for management effectiveness since the evaluation has been carried out

Protected Areas in Selected Arab Countries of the Levant Region (Syria, Lebanon & Jordan): An Evaluation of
Management and Recommendations for Improvement

33

previously (RSCN, 2008) using the METT tool, and the 2008 evaluation results created the opportunity for managing institutions (including RSCN) to apply adaptive management approaches and improve their management effectiveness results by 2011 (RSCN, 2008). Although not directly comparable, the results of METT evaluations in Jordan in 2008 had already shown positive results with no major deficiencies in management effectiveness of PAs studied, which implies that our evaluation provides an update, and confirms their strong performance (RSCN, 2008). The limitation of our results for Jordan is that two established PAs managed by ASEZA were not part of the analysis; however the absence of these areas is estimated to have minimally impacted the general evaluation, as they had comparable scores to the other Jordanian PAs in 2008 (RSCN, 2008).

Lebanon shows an interesting range of scores which can be attributed to several factors. This includes the fact that Lebanon has a special form of management of Nature Reserves which falls under the general jurisdiction of the MOE, while actual management effectiveness depends mostly on direct management capacities and resources. This is because the MOE acts mostly as a governance and administrative centre for Nature Reserves and has a very limited budget for reserve management (channeled through their respective managing institutions) that depends mostly on project (foreign) funding for developing management systems and improving effectiveness. Moreover, the sample taken from Lebanon is more varied than Jordan and Syria, as one biosphere reserve (Jabal Rihane) is a site protected by the Shiite Hezbollah party, and is a unique case of PA since it lacks any form of institutionalized management. The generally positive results obtained in Lebanon reflect local and national efforts of the MOE and mostly NGOs to protect valuable sites by persevering in attracting international funds and adhering to international program requirements (UNESCO MAB program) and standards.

Syrian PA management effectiveness scores showing results in the 'Basic' range can be misleading because of the very small sample size (3 of 27 established PAs). Since the responses were obtained specifically on the three most recognized PAs in Syria, which have benefited from the UNDP-GEF management development projects, and since Syria still lacks national management plans and effectiveness monitoring tools, the positive results obtained in our study for Syria could be an overestimate of management effectiveness across all 27 PAs. A more comprehensive research (including complementary quantitative data) on a larger and more representative sample would be needed to confirm the results obtained.

7.2 Which aspects of management are most effective?

According to our results, the most effective aspect of management is *planning*. This finding is consistent with the global survey results (Leverington at al., 2010), where the same *planning* indicators appeared in the top 7 scores, with the exception of 'Marking and security or fencing of park boundaries' which was the lowest scoring indicator in our study (Table 6). Syria and Lebanon have particularly noted this problem. In the case of Lebanon, this was reported earlier by Matar & Anthony (2010), and there appears to be a general absence of planning for this specific objective which may be due to complicated administrative and practical land tenure issues. Hence, for Lebanon and Syria, this problem might originate from the lack of national prioritization for this issue by local authorities governing the PAs (Ministries).

Planning indicators with particularly high scores include 'Protected area gazettal (legal establishment)', 'Appropriateness of design', 'Tenure issues', and 'Adequacy of protected area legislation and other legal controls'; which is consistent with the selection criterion for our study sample, i.e. "the presence of a national legal designation and/or international designation" (see Methods). Relatively weakly scoring indicators constitute *input* indicators, primarily those related to funding and staffing constraints, and 'adequacy of law enforcement capacity' (*process*), as well as 'external political and civil environment' constraints (*context*). This is likely due to the instability of funding for PAs, where there is high reliance on external financing institutions that have local/regional agendas and specific budgets under their agendas/programmes of work. This translates into an absence of highly-skilled persons being attracted to PA management positions which, in turn, decreases capacity for effective management including law enforcement. The low capacity of law enforcement at the managing institution level is also exacerbated by a generally weak enforcement at the national level reported in Lebanon (MOE-L et al., 2009).

7.3 Which factors are most related to overall effectiveness and successful outcomes?

Individual headline indicators most strongly correlated to overall management effectiveness (as reflected in item-total correlations in Table 7), show interestingly few similarities with the global survey results (Leverington et al., 2010). Only 'Effectiveness of administration including financial management' and 'Adequacy of infrastructure, equipment and facilities' scored in the top 5 of both studies. Indeed, 2 of our top 3 indicators ('Adequacy of human resource policies and procedures' and 'Conservation of nominated values-condition'), are not even included in the list of indicators in the international evaluation with $R>0.5$; our second most highly correlated outcome 'Appropriate program of community assistance' is only 19th in the list of indicators most correlated to outcomes globally as per Leverington et al. (2010). This highlights strengths that may be specific to the Levant region and could be developed and leveraged in the future. They could also provide interesting case-studies for the rest of the region (and others), to be more closely studied in the perspective of providing learning experiences for countries or regions that perform poorly in these areas/indicators.

Second, only one of the five most highly correlated indicators to either of the *outcomes* in our study scored highly in the global study, i.e. 'achievement of set work program', which was highly correlated with the 'conservation of values' *outcomes* indicator. The disparity observed between regional and global results provides an interesting case for further research in order to gain a deeper understanding of the relationship between these two *outcomes* indicators and overall management performance in specific PAs.

Finally, we found that those sites which were designated earlier had higher mean management effectiveness scores, a result which is consistent with the global study (Leverington et al., 2008). Since PAs of the Levant region are at various stages of designation and evolution, older PAs with more resource availability have had the time and capacity to develop, implement, and monitor their management plans, while others are still drafting them or planning to do so. This demonstrates the necessity to conduct regular PAME assessments to track effectiveness levels at various development stages within an individual site, and/or group of sites.

8. Conclusion

8.1 Research outcomes and contribution

This research is innovative as it provides the first PA management effectiveness evaluation on a regional scale in the Arab and Levant region and the first performance evaluation using the recently developed set of indicators by Leverington et al. (2010). Our results interestingly show a better than average performance score than the global results, with a remarkable 50% of surveyed PAs from Syria, Lebanon and Jordan scoring in the 'Sound management' range. Although this high score is mainly driven by consistently high scores in Jordan, Lebanon has also shown positive results and the small sample size of Syria consistently scored in the 'Basic' management range despite the lack of national monitoring strategies and action plans for systematic PAs management effectiveness evaluations in both Syria (SAR et al., 2009) and Lebanon (Matar & Anthony, 2010).

It is important to keep in mind that other factors – not studied in this research - can greatly affect the management performance difference between Jordan on one side, and Lebanon and Syria on the other, i.e. local political and economic stability. Lebanon has been on a long track of political instability, and was more recently shaken by the intense 2006 war against Israel; while Syria is currently witnessing a revolutionary transition that is dramatically destabilizing the country. This is an important factor that could be the subject of another study on the impact of national security and political stability on PAs management performance. Hanson et al. (2009), in their global review of warfare within biodiversity hotspots, point out that armed conflict often plays out in remote areas, and can lead to direct effects including ineffectiveness of PA boundaries, the withdrawal of PA staff, suspension of conservation activities, and an increase in uncontrolled hunting and grazing, the latter of which has already been identified in Lebanon (Matar & Anthony, 2010); MOE-L et al., 2009). Moreover, highlighted indirect effects include the emphasis on military spending at the expense of natural resource management. Our studied region is not immune to these effects and the baseline logic behind it is that the stability in the Hashemite Kingdom of Jordan creates a better enabling environment for managing institutions (RSCN and ASEZA) to advance and develop their PAs frameworks and performance.

8.2 Research limitations

Not unlike similar studies where respondent scoring is utilized to ascertain data on management effectiveness, our study is admittedly limited by the *subjectivity* of our respondents (Cook & Hockings, 2011). We have made every attempt to collect data from those respondents whom we believed had the best knowledge of the management indicators we were assessing, and with the *lack of published information in the region* (either qualitative or quantitative), this is a factor which we could not control for and which may be liable to overstating (or understating) performance by the individual assessors (Burgman, 2001). Moreover, our *PA selection criteria* deliberately excluded those sites which are under some level of 'protection', but are not formally recognized Reserves (e.g. *himas* in Lebanon, *rangelands* in Syria), which also limits the comparability with the global results of Leverington et al. (2010). Nonetheless, we use Leverington et al.'s study as the only available benchmark by which to make some comparisons on the effectiveness of our region to the global scene. Further, our results are consistent with the only similar studies or assessments which have been conducted in Lebanon (Matar & Anthony, 2010) and Jordan (RSCN, 2008), and issues identified in national reporting to the CBD by all three countries.

The above limitations are also exacerbated by simply the *lack of qualitative data* on management effectiveness in the region. This would be best addressed by conducting lengthy interviews or workshops with PA management staff, but was outside the scope of our study. This was compounded by the relatively low response rate from the Syrian PAs which, in all likelihood, distorts the national picture, particularly as our three responses were from PAs that have had the benefit of developed management structures (SAR et al., 2009). Without delving into the opinions of management staff on what obstacles or opportunities influence the effectiveness of the various indicators, we are limited in our analyses.

Despite these limitations, however, our study does provide a rapid and useful assessment of the management effectiveness of 18 PAs in Jordan, Lebanon, and Syria, and offers a platform for further research on this topic in the region.

8.3 Recommendations

With the aim of making a concrete contribution to the conservation field in the countries studied, and to address the need to draw regional lessons from PAME studies (Hockings et al., 2006), we recommend the following:

1. *Develop and adopt adapted management effectiveness evaluation tools that are based on the 6 evaluative elements, and integrate them into monitoring programmes for PAs in Syria and Lebanon by their respective Ministries of Environment (and/or other responsible governance institutions).* As Hockings et al. (2006: 48) recommended, "Evaluation of management effectiveness should be incorporated into the core business of protected area agencies." In the case of the Levant region, the superior performance of Jordanian PAs relative to Lebanon and Syria could be partly attributed to an already existing (and implemented) effective monitoring tool in Jordan. We believe that this finding could eventually provide an incentive for Lebanese and Syrian Authorities to start implementing a similar internationally recognized and standard monitoring tool for evaluation of management performance for the PAs under their jurisdictions. Our research results confirm other studies' findings, which suggest that comprehensive evaluations based on the WCPA Framework (the 6 elements) such as the one performed in Jordan, (i) provide a good overview of strengths and weaknesses of individual PAs, (ii) help identify management gaps, and (iii) can lead to more realistic recommendations and adaptive management actions to make improvements in the system (RSCN, 2008). This is one realistic recommendation that our research advances, since the implementation of such tools does not appear to be outside the scope of local institutional capacities (Matar & Anthony, 2010).

2. *Complement PAs national monitoring strategies with appropriate policies at the central decision-making institutional level.* For Syria, Lebanon and Jordan this would help to consolidate their implementation and ensure enforcement of policy by local authorities. As reflected in our study results, political support was part of the weakest scoring indicators that needs to be improved in the region. Creating and enforcing policies that would make management effectiveness evaluations a 'requirement' could be one avenue for local governments to address this issue, provided that this would be accompanied by capacity-building for managing institutions to carry out this obligation.

3. *Develop and adopt management effectiveness evaluation plans and monitoring programmes at individual PA levels for PAs with independent management systems (where existing national PA management strategies don't apply).* For example, in the case where Biosphere Reserves

with no legal national designation are independently managed by NGOs not governed by national monitoring strategies, it is advisable that these institutions develop their 'own' management evaluation plans and long-term monitoring strategies until the reserves acquire a national legal structure. Ideally, for continuity and more seamless transitions in implementation by different management structures, these would be based on existing and standardized evaluative tools, such as the ones based on the IUCN-WCPA Framework. Moreover, as Hockings et al. (2006: 49) state: "Evaluations that are integrated into the managing agency's culture and processes are more successful and effective in improving management performance in the long-term." Hence, there is value in adapting and integrating the chosen evaluation plans to existing management structures and processes.

4. *Consolidate the management structure and capacities of Biosphere Reserves by assigning a nationally recognized legal structure/designation.* This process would entitle Biosphere Reserves to local managerial arrangements, and align them with local policy requirements. As emphasized by Stoll-Kleemann et al. (2008: 11), "In order to effectively manage and conserve biodiversity in-situ, protected areas must be legally established, and management actions and measures must be implemented."

5. *Increase cooperation and networking between Lebanon, Syria, and Jordan for sharing experiences and learning best practices on PA management and monitoring.* This recommendation is aligned with Hockings et al. (2006: 49) call to "... learn from others and use or adapt existing methodologies if possible." The successful use of the METT tool in Jordan in 2008 reflects the existence of know-how and required skills in this region for the implementation of such tools. Hence, increased cooperation and sharing of experiences could foster the required transfer of skills and knowledge for successful implementation in Syria and Lebanon as well. Management effectiveness evaluations are not a 'one-time' process and need to be integrated into an overall management system where they would ideally be implemented on a regular basis, providing useful feedback for an effective overall adaptive management approach (Salafsky et al., 2001). Hence, we recommend that Jordan continues implementing (and adapting) the METT tool on a regular basis to monitor trends in its PAs management effectiveness; while Syria and Lebanon plan for regular evaluations when related strategies are developed and implemented.

9. Acknowledgment

We thank the CEU Department of Environmental Sciences and Policy CENSE research centre for funding; Ministries of Environment in Lebanon and Syria, Royal Society for the Conservation of Nature (RSCN) in Jordan, and individuals from PAs management teams who contributed to this research. We thank Viktor Lagutov for technical support, and Sylvia Abonyi, Ghassan Ramadan Jaradi, and the book editor for comments on an earlier draft of this manuscript.

10. References

2010 Biodiversity Indicators Partnership. (2010). *Biodiversity Indicators and the 2010 Target: Experiences and Lessons Learnt from the 2010 Biodiversity Indicators Partnership.* Secretariat of the Convention on Biological Diversity, ISBN 9292252720, Montreal, Canada

Anthony, B. (2008). Use of Modified Threat Reduction Assessments to Estimate Success of Conservation Measures within and Adjacent to Kruger National Park, South Africa. *Conservation Biology*, Vol.22, No.6, pp. 1497-1505, *ISSN 0888-8892*

Anthony, B.P. & Szabo, A. (2011). Protected Areas: Conservation Cornerstones or Paradoxes? Insights from Human-Wildlife Conflicts in Africa and Southeastern Europe. In: *The Importance of Biological Interactions in the Study of Biodiversity*. López-Pujol, J. (Ed.), pp. 255-282, InTech, ISBN 978-953-307-751-2, Rijeka, Croatia

Bonham, C. A.; Sacayon, E. & Tzi, E. (2008). Protecting imperiled paper parks: potential lessons from the Sierra Chinajá, Guatemala. *Biodiversity and Conservation*, Vol.17, No.7, pp. 1581-1593, ISSN 0960-3115

Butchart, S.H.M.; Walpole, M.; Collen, B.; van Strein, A.; Scharlemann, J.P.W.; Almond, R.E.A.; Baillie, J.; Bomhard, B.; Brown, C.; Bruno, J.; Carpenter, K.; Carr, G.M.; Chanson, J.; Chenery, C.; Csirke, J.; Davidson, N.C.; Dentener, F.; Foster, M.; Galli, A.; Galloway, J.N.; Genovesi, P.; Gregory, R.; Hockings, M.; Kapos, V.; Lamarque, J-F.; Leverington, F.; Loh, J.; McGeogh, M.; McRae, L.; Minasyan, A.; Morcillo, M.H.; Oldfield, T.; Pauly, D.; Quader, S.; Revenga C.; Sauer, J.; Skolnik, B.; Spear, D.; Stanwell-Smith, D.; Symes, A.; Spear, D.; Stuart, S.; Tyrrell, T.D.; Vie, J.C. & Watson, R. (2010). Global Biodiversity: Indicators of Recent Declines. *Science*, Vol.328, No.5982, (May 2010), pp. 1164-1168, ISSN 0036-8075

Burgman, M.A. (2001). Flaws in subjective assessments of ecological risks and means for correcting them. *Australian Journal of Environmental Management*, Vol.8, No.4, pp. 219-226, ISSN 1322-1698

Cantu-Salazar, L. & Gaston, K.J. (2010). Very Large Protected Areas and their Contribution to Terrestrial Biological Conservation. *Bioscience*, Vol.60, No.10, pp. 808-818, ISSN 0006-3568

Chape, S.; Harrison, J.; Spalding, M. & Lysenko, I. (2005). Measuring the Extent and Effectiveness of Protected Areas as an Indicator for Meeting Global Biodiversity Targets. *Philosophical Transactions of the Royal Society B: Biological Sciences*, Vol.360, No.1454, (February 2005), pp. 443-455, ISSN 1471-2970

Coad, L.; Burgess, N.; Fish, L.; Ravillious, C.; Corrigan, C.; Pavese, H.; Granziera, A. & Besançon, C. (2008a). Progress towards the Convention on Biological Diversity Terrestrial 2010 and Marine 2012 Targets for Protected Area Coverage. *Parks*, Vol.17, No.2, (December 2008), pp. 35-42, ISSN 0960-233X

Coad, L.; Corrigan, C.; Campbell, A.; Granziera, A.; Burgess, N.; Fish, L.; Ravilious, C.; Mills, C.; Miles, L.; Kershaw, F.; Lysenko, I.; Pavese, H. & Besançon, C. (2008b). *State of the World's Protected areas Areas 2007: an Annual Review of Global Conservation Progress*. UNEP-WCMC, Cambridge, UK

Conservation International (CI). (2007). Official website. 27 May 2011, Available from: http://www.biodiversityhotspots.org/xp/hotspots/mediterranean/Pages/defaul t.aspx

Convention on Biological Diversity (CBD). (2010). Conference of the Parties (COP) 10, Decision X/31. *Protected Areas Section* 19(a). 22 October 2011, Available from: http://www.cbd.int/decision/cop/?id = 12297

Cook, C.N. & Hockings, C. (2011). Opportunities for Improving the Rigor of Management Effectiveness Evaluations in Protected Areas. *Conservation Letters*, Vol.4, No.5, (October-November 2011), pp. 372-382, ISSN 1755-263X

Craigie, I.D.; Baillie, J.E.M.; Balmford, A.; Carbone, C.; Collen, B.; Green, R.E. & Hutton, J.M. (2010). Large mammal population declines in Africa's protected areas. *Biological Conservation*, Vol.143, No.9, (September 2010), pp. 2221-2228, ISSN 0006-3207

Cuttelod, A.; García, N.; Abdul Malak, D.; Temple, H. & Katariya, V. (2008). The Mediterranean: a Biodiversity Hotspot Under Threat. In: *Wildlife in a Changing World: An Analysis of the 2008 IUCN Red List of Threatened Species*, J.-C. Vié, C. Hilton-Taylor and S.N. Stuart (eds), pp. 89-101, IUCN, ISBN 978-2-8317-1063-1, Gland, Switzerland

Dirzo, R. & Raven, P.H. (2003). Global State of Biodiversity and Loss. *Annual Review of Environment & Resources*, Vol.28, (November 2003), pp. 137-167, ISSN 1543-5938

Elzinga, C.L.; Salzer D.W.; Willoughby, & J.W. & Gibbs, J.P. (2001). *Monitoring Plant and Animal Populations*. Wiley-Blackwell, ISBN 978-0632044429, Abingdon, UK

Ervin, J. (2003). *Rapid Assessment and Prioritization of Protected Area Management (RAPPAM)*. WWF International, Gland, Switzerland

Gaston, K. J.; Jackson, S. F.; Cantú-Salazar, L. & Cruz-Pi͂nón, G. (2008). The ecological performance of protected areas. *Annual Review of Ecology, Evolution, and Systematics*, Vol.39, (December 2008) pp. 93–113, ISSN 1543-592X

Hagen, R. & Gerard, J. (2004). *Stable Institutional Structure for Protected Areas Management in Lebanon evaluation and recommendations*. Report prepared for the Ministry of Environment, Ministry of Environment, Lebanon

Hanson T.; Brooks, T.M.; da Fonseca, G.A.B.; Hoffmann, M.; Lameroux, J.F.; Machlis, G.; Mittermeier, C.G; Mittermeier, R.A. & Pilgrim, J.D. (2009). Warfare in Biodiversity Hotspots. *Conservation Biology*, Vol.23, No.3, pp. 578-587, *ISSN 0888-8892*

Hockings, M. & Phillips, A. (1999). How well are we doing? – some thoughts on the effectiveness of protected areas. *Parks*, Vol.9, No.2, (June 1999), pp. 5-14, ISSN 0960-233X

Hockings, C. S.; Solton, S. & Dudley, N. (2000). *Evaluating Effectiveness: a Framework for Assessing the Management of Protected Areas*. IUCN, ISBN 2-8317-0546-0, Gland, Switzerland and Cambridge, UK

Hockings, M. (2003). Systems for assessing the effectiveness of management in protected areas. *BioScience*, Vol.53, No.9, (September 2003), pp. 823-832, ISSN 0006-3568.

Hockings, M.; Stolton, S.; Leverington, F.; Dudley, N. & Courrau, J. (2006). *Evaluating Effectiveness: A framework for assessing management effectiveness of protected areas*. 2nd edition. IUCN, Gland, Switzerland & Cambridge, UK. ISBN 2-8317-0939-3

Holling, C.S. (1978). *Adaptive Environmental Assessment and Management*. John Wiley & Sons, ISBN 978-1932846072, New York

International Union for the Conservation of Nature and Natural Resources (IUCN). (2005). *The Durban Action Plan*. Revised version, March 2004. 21 October 2011, Available from: http://cmsdata.iucn.org/downloads/durbanactionen.pdf

International Union for the Conservation of Nature and Natural Resources (IUCN) - World Commission on Protected Areas (WCPA). (2009). WCPA *Science and Management Strategic Direction. Management Effectiveness as a Priority*. 2 March 2009, Available from: http://www.iucn.org/about/union/commissions/wcpa/wcpa_work/wcpa_strat egic/wcpa_science/

International Union for the Conservation of Nature and Natural Resources (IUCN) (2011). *The IUCN Red List of Threatened Species Version 2011.1*, 15 October 2011, Available from http://www.iucnredlist.org

Leverington, F.; Hockings, M.; Pavese, H.; Costa, K.L. & Courrau, J. (2008). *Management effectiveness evaluation in protected areas – A global study. Supplementary Report No. 1.Overview of approaches and methodologies.* The University of Queensland, TNC, WWF, & IUCN-WCPA, Gatton, Australia

Leverington, F.; Costa, K.L.; Pavese, H.; Lisle, A. & Hockings, M. (2010). A Global Analysis of Protected Area Management Effectiveness. *Environmental Management,* Vol.46, pp. 685-698, ISSN 0364-152X

Living University. (2009). *The Levant.* 21 October 2011. Available from: http://bibarch.com/ArchaeologicalSites/index.htm

MacKinnon, J.; MacKinnon, K.; Child, G. & Thorsell, J. (1986). *Managing Protected Areas in the Tropics,* IUCN, ISBN 978-2880328085, Cambridge, UK

Margules, C.R. & Pressey, R.L. (2000). Systematic Conservation Planning. *Nature,* Vol.405, (May 2000), pp. 243-253, ISSN 0028-0836

Matar, D.A. & Anthony, B.P. (2010). Application of Modified Threat Reduction Assessments in Lebanon. *Conservation Biology,* Vol.24, No.5, pp. 1174–1181, *ISSN 0888-8892*

Medail, F. & Quezel, P. (1997). Hot-spots Analysis for Conservation of Plant Biodiversity in the Mediterranean Basin. *Annals of the Missouri Botanical Garden,* Vol.84 , No.1, pp.112-127, ISSN 0026-6493

Medail, F. & Quezel, P. (1999). Biodiversity Hotspots in the Mediterranean Basin: Setting Global Conservation Priorities. *Conservation Biology,* Vol.13, No.6, pp. 1510-1513, *ISSN 0888-8892*

Ministry of Environment in Jordan (MOE-J). (2009). *Fourth National Report for Jordan to the Convention on Biological Diversity.* Ministry of Environment, Amman, Jordan

Ministry of Environment in Lebanon (MOE-L). (2002). *Biological Diversity Second National Report to Conference of the Parties,* Ministry of Environment, Lebanon

Ministry of Environment in Lebanon (MOE-L). (2005). *Stable Institutional Structure for Protected Areas Management in Lebanon: Towards a Stable Institutional Management Structure.* Report prepared by EcoDit Liban for the Ministry of Environment, Lebanon

Ministry of Environment in Lebanon (MOE-L). (2006a). *Stable Institutional Structure for Protected Areas Management in Lebanon: Monitoring and evaluation indicators for protected areas.* Report produced by EcoDit Liban, Ministry of Environment, Lebanon

Ministry of Environment in Lebanon (MOE-L). (2006b). *Stable Institutional Structure for Protected Areas Management in Lebanon: Protected Areas Categories Report.* Unpublished report produced by EcoDit Liban (used with permission), Ministry of Environment, Lebanon

Ministry of Environment in Lebanon (MOE-L), Global Environment Facility (GEF), & United Nations Development Programme (UNDP). (2009). *Fourth National Report of Lebanon to the Convention on Biological Diversity,* Ministry of Environment, Lebanon

Ministry of Environment in Lebanon (MOE-L) & Lebanese University (LU). (2004a). *Biodiversity Assessment and Monitoring in the Protected Areas/Lebanon LEB/95/G31. Final Report. Horsh Ehden Nature Reserve.* Report prepared for the Ministry of the Environment. Lebanon: Ministry of Environment. 2 March 2009, Available from: http://biodiversity.moe.gov.lb/

Ministry of Environment in Lebanon (MOE-L), & Lebanese University (LU). (2004b). *Biodiversity Assessment and Monitoring in the Protected Areas/Lebanon LEB/95/G31. Final Report. Al-Shouf Cedar Nature Reserve.* Report prepared for the Ministry of the

Environment. Ministry of Environment, Lebanon, 10 March 2009, Available from: http://biodiversity.moe.gov.lb/

Ministry of Environment (MOE-L), United Nations Development Program (UNDP), & ECODIT. (2011). *State and Trends of the Lebanese Environment 2010.* Prepared by ECODIT for the MOE and UNDP

Mittermeier, R.A.; Gil, P.R.; Hoffmann, M.; Pilgrim, J.; Brooks, T.; Mittermeier, C.G.; Lamoreux, J. & da Fonseca, G.A.B. (2004). *Hotspots Revisited.* Cemex, ISBN 9686397779 , Mexico City, Mexico

Mulongoy, K.J. & Chape, S. (2004). *Protected Areas and Bodiversity: An Overview of Key Issues. UNEP-WCMC Biodiversity Series (21).* CBD and UNEP-WCMC, ISBN 92 804 2404 5, Cambridge, UK

Myers, N.; Mittermeier, R.A.; Mittermeier, C.G.; da Fonseca, G.A.B. & Kent, J. (2000). Biodiversity Hotspots for Conservation Priorities. *Nature*, Vol.403, (February 2000), pp. 853–858, ISSN 0028-0836

Oates, J.F. (1999). *Myth and Reality in the Rain Forest: How Conservation Strategies Are Failing in West Africa.* University of California Press, ISBN 0-520-21782-9, Berkeley and Los Angeles

Persha, L., & Rodgers, A. (2002). Threat Reduction Assessment in the UNDP-GEF East African Cross Borders Biodiversity Project: Experience with a New ICD Monitoring Tool. *ArcJournal*, Vol.14, (August 2002), Tanzania Forest Conservation Group, Dar es Salaam, Tanzania

Rodrigues, A.S.L.; Andelman, S.J. ; Bakarr, M.I.; Boitani, L.; Brooks, T.M.; Cowling, R.M.; Fishpool, L.D.C.; da Fonseca, G.A. B.; Gaston, K.J.; Hoffmann, M.; Long, J.S.; Marquet, P.A.; Pilgrim, J.D.; Pressey, R.L.; Schipper, J.; Sechrest, W.; Stuart, S.N.; Underhill, L.G.; Waller, R.W.; Watts, M.E. J. & Yan, X. (2004). Effectiveness of the Global Protected Area Network in Representing Species Diversity. *Nature*, Vol.428, No.6983, pp. 640-643, ISSN 0028-0836

Royal Society for the Conservation of Nature (RSCN). 2008. *Jordan Protected Areas: Management Effectiveness. National Report 2008.* RSCN, Amman, Jordan

Sabatinelli, G. (2008). *The Scarabs of the Levant: Syria, Lebanon, Jordan, Palestine, Israel, and Sinai.* 21 October 2011. Available from: http://www.glaphyridae.com/Biogeografia/NEL.html

Salafsky, N., & Margoluis, R. (1999). Threat Reduction Assessment: a Practical and Cost Effective Approach to Evaluating Conservation and Development Projects. *Conservation Biology*, Vol.13, No.4, (August 1999), pp. 830-841 *ISSN 0888-8892*

Salafsky, N.; Margoluis, R. & Redford, K.H. (2001). *Adaptive management: a tool for conservation practitioners.* Biodiversity Support Program, Washington, D.C.

Salafsky, N.; Margoluis R.; Redford, K.H. & Robinson, J.B. (2002). Improving the Practice of Conservation: a Conceptual Framework and Research Agenda for Conservation Science. *Conservation Biology*, Vol.16, No.6, (December 2002), pp. 1469-1479, *ISSN 0888-8892*

Scheffé, H. (1953). A method for judging all contrasts in the analysis of variance. *Biometrika*, Vol. 40, No.1-2, pp. 87-104, ISSN 0006-3444

Secretariat of the Convention on Biological Diversity (SCBD). (2009). *Protected Areas.* 2 March 2009, Available from: http://www.cbd.int/protected

Sodhi, N.S.; Butler, R.; Laurance, W.F. & Gibson, L. (2011). Conservation successes at micro-, meso- and macroscales. *Trends in Ecology and Evolution*, Vol.26, No.11 (November 2011), pp. 585-594, ISSN 0169-5347

Stoll-Kleemann, S.; Bertzky, M.; de la Vega-Leinert, A. C.; Fritz-Vietta, N.; Leiner, N.; Hirschnitz-Garbers, M.; Mehring, M.; Reinhold, T. & Schliep, R. (2008). *The Governance of Biodiversity (GoBi) Project: A Vision for Protected Area Management and Governance.* Ernst-Moritz-Arndt-Universität Greifswald, Germany

Syrian Arab Republic (SAR), Global Environment Facility (GEF), & United Nations Development Programme (UNDP). (2009). *The Fourth National Report on Biodiversity in the Syrian Arab Republic.* Report prepared for the Ministry of Environmental Affairs, Syria

Talhouk, N.S. & Abboud, M. (2009). Impact of Climate Change: Vulnerability and Adaptation - Ecosystems and Biodiversity, In: *Arab Environment: Climate Change-Impact of Climate Change on Arab Countries,* K. T. Mostafa & N.W. Saab, (Eds), 101-112, Arab Forum for Environment and Development. ISBN: 9953-437-28-9, Beirut, Lebanon

Tucker, G. (2005). *A Review of Biodiversity Conservation Performance Measures.* Earthwatch Institute, Oxford, UK

United Nations Development Programme (UNDP). (1995). *Project Document: Protected Areas for Sustainable Development (Protected Areas Project).* Available at Green Line Association public library

United Nations Development Programme (UNDP). (2004). *Biodiversity Conservation and Management: Project Document.* 1 June 2011, Available from: http://www.undp.org.sy/files/227BiodiversityConservationandProtectedAreaManagement.pdf

United Nations Development Programme (UNDP). (2005). *Third National Report for Lebanon to the Convention on Biological Diversity.* UNDP, Lebanon

United Nations Environment Programme (UNEP). (2006). Conference of the Parties on the Convention for Biological Diversity. Eighth meeting held in Curitiba, Brazil, 20-31 March 2006: Item 27.1 of the provisional agenda. Review of the implementation of the programme of work on protected areas for the period 2004-2006. 1 March 2009, Available from: http://www.cbd.int/doc/meetings/cop/cop-08/official/cop-08-29-en.pdf

United Nations Educational Scientific and Cultural Organization (UNESCO). (2011). *Directory of the World Network of Biosphere Reserves (WNBR).* 23 October 2011, Available from: http://www.unesco.org/new/en/natural-sciences/environment/ecological-sciences/biosphere-reserves/world-network-wnbr/wnbr/

Weaver D. B. (2001). *The Encyclopedia for Ecotourism.* CAB International, ISBN 978-0851996820 Oxon, UK.

World Wildlife Fund (WWF). (2007). *Management Effectiveness Tracking Tool: Reporting Progress at Protected Area Sites.* Second edition (July 2007). World Wildlife Fund, Gland, Switzerland

World Wildlife Fund (WWF) & World Bank (WB) (2003) (revised in 2005). *Reporting Progress at Protected Area Sites: a Simple Site-level Tracking Tool Developed for the World Bank and WWF.* World Wildlife Fund, Gland, Switzerland, 10 March 2009, Available from: http://assets.panda.org/downloads/patrackingtool.pdf

3

Introgression and Long-Term Naturalization of Archaeophytes into Native Plants – Underestimated Risk of Hybrids

Hiroyuki Iketani[1] and Hironori Katayama[2]
[1]Institute of Fruit Science, National Agriculture and Food Research Organization,
Food Resources Education and Research Center,
[2]Graduate School of Agricultural Science, Kobe University,
Japan

1. Introduction

1.1 Difficulties in the study of archaeophytes

Alien plants pose problems in the conservation of biodiversity, especially by invasion and successive mal-effects on local ecosystem and biodiversity (Pyšek et al., 1995; Ellstrand, 2003; Nentwig, 2007). Even if they are only in cultivation, they could affect by hybridization, vector of diseases and pests and other factors. Since an alien plant is defined as one whose distribution has expanded out of its native range under human influence, the history of alien plants begins with human migration, especially in association with agriculture. As most have spread with active migration since the Age of Discovery, alien plants can be delineated into archaeophytes (introduced before the Age of Discovery) and neophytes (introduced since) (Pyšek et al., 2002; Preston et al., 2004).

Although this classification is commonly used, particularly in Europe, it is difficult to distinguish these two categories (Pyšek et al., 2004; Willis & Birks, 2006). It is practically impossible to prove the non-nativeness of a plant, especially from morphological, ecological, or phytogeographical data. For example, most archaeophytes grow in or near human-made environments, for example, as agricultural weeds or ruderal plants. They have already extended into that ecological system and have been held in equilibrium. Therefore, although research tries to distinguish archaeophytes from true native plants, the two are usually treated the same in the practice of biodiversity conservation. In fact, some archaeophytes are listed in Red Lists (Cheffings & Farrell, 2005; Ministry of Environment, Japan, 2007), since exclusion of archaeophytes from native plants without distinct evidence would make their conservation value lower (Willis & Birks, 2006).

1.2 Biodiversity in Japan and archaeophytes

The Japanese archipelago is rich in biodiversity (Ohwi, 1965; Iwatsuki et al., 1993–2011). The climate of this region ranges widely from subarctic to subtropical, with high humidity and precipitation throughout. It lies near the Asian continent and has been connected with it several times in geological history (continental islands). This placement may have promoted migration from north and south and produced flora with mixed boreal and temperate elements. By contrast, the archipelago was not connected with the continent during the last glacial maximum, and this isolation may have stimulated the phylogenetic differentiation of species or intraspecific taxa from the continental mother taxa. In addition, its mountainous topography with about twenty 3000-m-class peaks provides refugia for many alpine plants that are relicts of the glacial period.

However, this biologically affluent archipelago is also one of the most populous regions in the world. Although preservation measures beginning in the early modern age have maintained a high percentage of forest (about 70% of land) and protected this "green archipelago" (Totman, 1987), population pressure still poses threats. Therefore Japan is nominated in one of the biodiversity hotspots (Mittermeier et al., 2004).

Humans arrived in Japan relatively recently, ca. 35 000 years BP or later (Keally, 2009). Agriculture is estimated to have begun in Japan 3000 years BP or later (Shōda, 2007). The introduction of alien plants may have begun from this period. Maekawa (1943) first proposed the concept of "prehistorically naturalized plants", presumably introduced from the Asian continent before the beginning of written history (end of the 6th century CE). He listed about 120 ruderal plants, mainly agricultural weeds. Other plants, including trees, are now also included in this category (Shimizu, 2003). Of course, many plants were also introduced from the Asian continent in historic times before the Age of Discovery. However, since most of these plants are crops or ornamentals, only a few became completely naturalized. Therefore, most of the naturalized archaeophytes in Japan are prehistorically naturalized plants.

1.3 Archaeophytic naturalized useful trees in Japan

Several fruit trees and a few industrial crop trees were introduced into Japan prehistorically. Representative examples are pear (*Pyrus pyrifolia* Nakai), peach (*Prunus persica* L.), plum (*Prunus salicina* L.), apricot (*Prunus armeniaca* L.), persimmon (*Diospyros kaki* Thunb.), loquat (*Eriobotrya japonica* Lindl.), and princess tree (*Paulownia tomentosa* Steud.). These plants now grow more or less wild and can reproduce without human intervention. Some have extended their wild distribution into the upper temperate deciduous forest zone (1000–1600 m a.s.l. in central Japan), where human settlement has been rare and many native flagship species of Japanese biodiversity grow. Whereas some or all of these species were formerly regarded as native, archaeological and phytogeographical research now estimates that they are introduced; for example, the remains, usually seeds or stones, of these fruit trees are found only at archaeological sites dating after the beginning of agriculture, in contrast to native edible fruit species such as wild grape (*Vitis coignetiae* Planch.) and raspberries (*Rubus* spp.) (Kobayashi, 1990).

2. Introgression of archaeophytic *Pyrus pyrifolia* into native *Pyrus ussuriensis* in NE Japan

2.1 Problems in the classification of Japanese *Pyrus*

Among these naturalized useful trees, only *Pyrus pyrifolia* has intercrossable wild relatives in the flora of Japan, namely *Pyrus calleryana* and *Pyrus ussuriensis*. *Pyrus calleryana* is doubtlessly native because it has a very distinctive distribution in common with other famous endangered plants such as *Magnolia stellata* (Ueda, 1989). On the other hand, the status of *P. ussuriensis* in the flora of Japan has been obscure. This plant was originally described in the early 20th century as two new native species from remote regions (Nakai, 1918): namely, *Pyrus hondoensis* in central Japan (Chubu region) and *Pyrus aromatica* in northeastern Japan (Kitakami Mountains; Fig. 1). However, since then, trees morphologically intermediate between these plants and *P. pyrifolia* were reported, and many of them were described as distinct species. As a result, more than 70 "species" of *Pyrus* were described throughout Japan. Since most of them exist only sporadically in human areas, modern taxonomists treat them as synonyms of either *P. pyrifolia* or *P. ussuriensis* (Ohwi, 1965; Kitamura, 1979; Iketani & Ohashi, 2001).

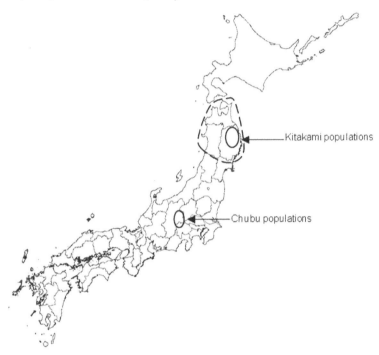

Solid ellipses indicate the distribution site of the populations in the Kitakami Mountains and the Chubu region. The dashed line indicates the region where wild trees other than the true native type are common. Prefectural boundaries are also shown.

Fig. 1. Localities of two native populations of *Pyrus ussuriensis* in Japan

This taxonomic confusion leads to doubt about the nativeness of wild populations of *P. ussuriensis* (Kitamura, 1979), which is heightened for other reasons. First, this species is edible and is cultivated in northeastern China and Korea. Therefore, *P. ussuriensis* could also have been introduced and naturalized. Second, species of *Pyrus* easily hybridize with each other. Third, morphological distinction between species is obscure: only the presence/absence of calyx lobes in mature fruits is the discriminative character in floristic and taxonomic works (Yü & Ku, 1974; Ku & Sponberg, 2003). Finally, the original sites of discovery are in the lower mountain zone, not far from human settlements.

2.2 Approaches from phytogeography and morphology

As one of the main reasons for this suspicion may be a lack of understanding of the true wild populations of *P. ussuriensis*, my group performed extensive field investigations (Iketani & Ohashi, 2003). We found this species both near the original places of discovery and in more elevated places, which correspond to the upper temperate deciduous forest zone. In contrast to the sporadic distribution of trees growing in human areas at lower elevation, the wild trees grew more densely at higher elevation. They were also morphologically distinct from other seemingly wild or cultivated trees. The fruits and leaf laminae are small in the former but vary in the latter from as small as in wild trees to as large as in edible cultivars (Fig. 2). In addition, the disjunctive geographical distribution at higher elevations of the Chubu region and the Kitakami Mountains is one of the distinctive patterns of endemic taxa in the flora of Japan (Ohashi, 1987). These findings support the indigenousness of the wild populations.

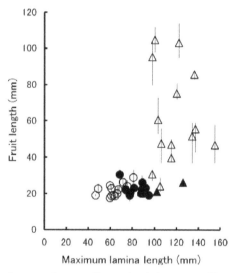

• *Pyrus ussuriensis* from northeastern Japan. ○ *P. ussuriensis* from central Japan. ▲ *P. ussuriensis* from Asian continent. △ Cultivars of *P. ussuriensis* and naturalized *Pyrus*. Vertical lines show the maximum and minimum values of fruit length. (From Iketani & Ohashi, 2003)

Fig. 2. Relation between fruit length and maximum lamina length in *Pyrus* accessions

2.3 Molecular approach reveals introgression between native and prehistorically naturalized plants

Traditional methods made the presence of true native populations plausible. However, the evidence was not conclusive, and the reason for the appearance of morphologically intermediate trees between *P. ussuriensis* and *P. pyrifolia* remained in doubt. Worse, many pear trees with wide morphological variations were discovered in the Kitakami Mountains and surrounding region (Katayama & Uematsu, 2006). To resolve these difficulties, we investigated population genetics using microsatellite loci and with Bayesian statistical inference (Iketani et al., 2010).

The analysis of 226 individuals from six regions implied five hypothetical ancestral populations (Fig. 3; Iketani et al., 2010). These results were at least partly predictable, but not entirely expected. The two true native populations of *P. ussuriensis* in Japan showed genetic distinctiveness, and both were much differentiated from wild plants of *P. ussuriensis* in the Asian continent. These results show that the Japanese populations of *P. ussuriensis* are truly native. Similarly, the introduction of *P. pyrifolia* from China was also supported, since the genetic structure of old Japanese cultivars and local landraces shows a partially common element with Chinese pear cultivars.

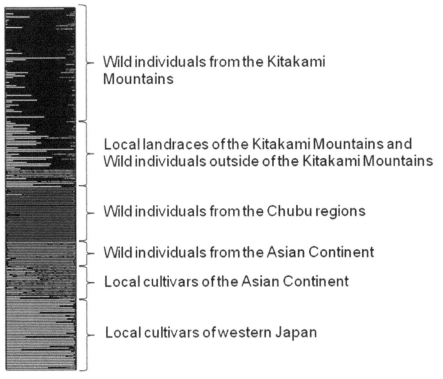

Wild individuals from the Kitakami Mountains

Local landraces of the Kitakami Mountains and Wild individuals outside of the Kitakami Mountains

Wild individuals from the Chubu regions

Wild individuals from the Asian Continent

Local cultivars of the Asian Continent

Local cultivars of western Japan

Five hypothetical ancestral populations are shown with different shadings. Modified based on Iketani et al., (2010)

Fig. 3. Bayesian statistical inference of population structure of *P. ussuriensis* and *P. pyrifolia*.

Populations from northeastern Japan were more or less genetically admixed with *P. pyrifolia*. This phenomenon was more conspicuous in local landraces and wild individuals collected from outside of the Kitakami Mountains, but was also evident in wild individuals in that region. This result and the ubiquity of intermediate trees clearly show introgression between native *P. ussuriensis* and prehistorically naturalized *P. pyrifolia* trees. In addition, truly native trees proved to be much rarer than introgressed trees and should be protected. *Pyrus ussuriensis* has now been added to the Japanese National Red List of Threatened Plants (Ministry of Environment, Japan, 2007).

3. Implications for biological conservation

3.1 Long-term hybridization and naturalization?

At present we have almost no empirical evidence to infer the historical introgression of *P. pyrifolia* into native *P. ussuriensis*. Although agriculture began later in northeastern Japan than in western Japan, the introduction of *P. pyrifolia* into this region would have also begun prehistorically, or early in historical times at the latest. Since *P. pyrifolia* itself is not widely escaped and naturalized even now, hybridization might have occurred first, and naturalization of hybrid offspring might have followed. Acceleration of naturalization due to hybridization between alien and native species, which is a well-known conservation problem (Ellstrand, 2003; Fitzpatrick & Shaffe, 2007), might have happened.

However, there are still many unclear points in the above scenario. Neither *P. ussuriensis* nor *P. pyrifolia* is very invasive. When assessed for invasiveness, they are judged as posing no or limited risk. For example, their weed risk factor (Food and Agriculture Organization of the United Nations [FAO], 2005) is 4 at most, against a critical score of 6. Even if the fitness of hybrids is better than that of the parents, it is unlikely that they would have expanded like other invasive plants. Therefore, we have to ascribe the cause of this limited expansion to the long passage of time, perhaps more than a thousand years. Kowarik (1995) and Pyšek & Jarošík (2005) studied the time-lag between the introduction and naturalization of alien plants in terms of centuries. The case of *Pyrus* could be appended as an example in terms of millennia.

A long residence time must raise the chance of successful naturalization (Pyšek & Jarošík, 2005). However, human activities such as repeated secondary release would promote invasion beyond the threshold of naturalization (Kowarik, 2003). As both *Pyrus* species bear edible fruits, ancient peoples in northeastern Japan might have grown and propagated hybrid pear trees, which would have become the source of secondary release, although there is no supportive evidence from history, ethnology, archaeology, or demography. Thus, research in these fields will be necessary.

3.2 Negative effect of archaeophytes through hybridization

How should we treat these hybrid pears? On one hand, they now grow widely in northeastern Japan but are not very invasive. The number of individuals is not very large and they are perhaps held in equilibrium in nature. Therefore, they could be treated similarly as native plants, as many other prehistorically naturalized plants (archaeophytes) are.

However, we cannot ignore these pears, because the plant that now grows widely is not an archaeophyte itself. Instead, an archaeophyte has hybridized with a native plant, and the hybrid offspring grow better than their native parent. In this case, hybrid plants should be controlled for the conservation of native plants, especially if the latter are threatened.

This case shows that archaeophytes that are themselves not invasive could still pose a risk to native plants in the long term. The negative effects of alien plants on natural ecosystems are already known. For example, there is much evidence that crops and other domesticated plants can hybridize with their wild relatives while still in cultivation (Ellstrand, 2003). Reproductive interference, which is the depression of fitness by interspecific pollination interactions such as competition for limited pollinators and pollen loss by interspecific pollen transfer, has recently been recognized (Matsumoto et al., 2010; Takakura et al., 2011). In both phenomena, negative effects could become evident on the human timescale if a plant was cultivated extensively, such as agricultural crops. Even with ornamental use, the effect could appear on such a timescale.

Sukopp & Sukopp (1993) expressed concern about the naturalization ("becoming feral") of cultivated plants ("cultigens") and long-term ecological effects such as hybridization with related wild plants. They listed several pairs of cultivated and wild plants in Central Europe which could hybridize. Since then, several actual cases among these pairs have been reported; e.g., *Daucus carota* ssp. *sativus* and ssp. *carota* (Magnussen & Hause, 2007), *Beta vulgaris* ssp. *vulgaris* and ssp. *maritima* (Arnauld et al., 2003; Fénart et al., 2008), and *Brassica napus* and *Brassica rapa* (Andersen et al., 2009).

In conservation policy, introduced cultivated plants which prove not to be invasive are not usually regard as dangerous. However, in the conservation of threatened plants, their crossable relatives should not be grown near natural habitat. The International Union for Conservation of Nature and Natural Resources [IUCN] (2000) recommended: "Since the impacts on biological diversity of many alien species are unpredictable, any intentional introductions and efforts to identify and prevent unintentional introductions should be based on the precautionary principle." The case of *Pyrus* supports this policy. We have to control even archaeophytes if they could harm native plants, once they prove to be truly naturalized alien plants. This is why we should distinguish archaeophytes and true native plants more precisely.

3.3 Potential problems in the Pyrinae

The invasiveness and other negative ecological effects of *Pyrus* and related plants are also relevant to the case for distinguishing archaeophytes and true native plants more precisely. The genus *Pyrus* belongs to the subtribe Pyrinae of the tribe Pyreae (formerly subfamily Maloideae) in the family Rosaceae (Potter et al., 2007). Since the reproductive barriers between the members of this tribe are low, interspecific hybrids which are usually fertile are common in nature; even intergeneric hybrids are not rare. This is one of the most extreme cases in higher plants, even though plants produce hybrids much more easily than animals. Previously this character has been a problem only in taxonomy (Kovanda, 1965; Robertson et al., 1991), but it could also become a problem in conservation. Species of this subtribe naturalize commonly in the temperate zones of both hemispheres; e.g., *Malus pumila* (apple), *Pyrus communis* (common pear), *P. pyrifolia, Cotoneaster* spp., and *Crataegus* spp. (hawthorn).

So far, only a few species of the Pyrinae are nominated in national and other lists of invasive plants (Table 1). This list shows the obvious invasiveness of some species in Oceania and the Pacific islands, where no native species of the Pyrinae grow (except for *Osteomeles*). *Eriobotrya japonica, Pyracantha* spp., and *Sorbus* spp. are also naturalized in the temperate zone of Europe and Japan (Tutin et al., 1968; Shimizu, 2003), but they are not recognized as invasive. However, hybridization between introduced plants and their wild relatives would happen in the long term even without naturalization.

Recently, the invasiveness of *Pyrus calleryana*, which was introduced from East Asia and has been widely propagated in the USA as an ornamental during the past 50 years, has been

Species	Nominated list[a]
Amelanchier spicata	EMPPO
Cotoneaster franchettii	AWC, CIPC
Cotoneaster glaucophyllus	AWC, MoE NZ
Cotoneaster lacteus	CIPC
Cotoneaster pannosus	AWC, HER, CIPC
Cotoneaster salicifolius	AWC
Cotoneaster simonsii	AWC, MoE NZ
Crataegus monogyna	AWC, CIPC
Crataegus sinaica	AWC
Eriobotrya japonica	MoE NZ, HER
Pyracantha angustifolia	AWC, HER, MoE NZ
Pyracantha coccinea	AWC
Pyracantha fortuneana	AWC
Sorbus aucuparia	MoE NZ
Sorbus spp.	AWC

a: AWC, Australian Weeds Committee (2011). CIPC, California Invasive Plant Council (2006). EMPPO, European and Mediterranean Plant Protection Organization (2011). HER, Hawaiian Ecosystems at Risk (2011). MoE NZ, Ministry of Environment, New Zealand (2011).

Table 1. Species of the Pyrinae nominated in national and other lists of invasive plants.

recognized (Vincent, 2005). This is perhaps the first case of a species of *Pyrus* being recognized as invasive. This sudden realization of invasion is explained as the overcoming of self-incompatibility owing to the planting of different cultivars and the Allee effect (Culley & Hardiman, 2009; Hardiman & Culley, 2010). However, since there are no native species of *Pyrus* in North America, negative genetic effects on native relatives would not occur.

In Europe, in contrast, although Sukopp & Sukopp (1993) stated that *Pyrus communis* became feral in Europe in the absence of wild relatives, in fact there are many wild species in this continent (Tutin et al., 1968; Aldasoro et al., 1996), some possibly of hybrid origin, either wild × wild or wild × cultivated. Introgression between cultivated apple (*Malus pumila*, syn. *M. ×domestica*) and a wild relative (*Malus sylvestris*) in Europe is already known (Coart et al., 2003, 2006; Larsen et al., 2006). These cases are comparable to our case of *Pyrus ussuriensis* in terms of introgression between archaeophytic cultivated fruit trees and wild relatives.

The most crucial but incompletely understood situation occurs in East Asia, the center of differentiation of both *Pyrus* and *Malus*, as well as of many other genera of the Pyrinae. About 20 species of *Pyrus* and 30 of *Malus* are now recognized in this region (Iketani & Ohashi, 2001; Ohashi, 1993; Wu et al., 2003). Some species, especially those now found only in cultivation, might have a hybrid origin. These hybridizations might have occurred among native species, although some might have occurred between alien and native. For example, the Chinese pear cultivars we studied proved to be admixtures between *P. ussuriensis* and *P. pyrifolia* (Fig. 3). These cultivars originated in northern China (Yü, 1979), where *P. ussuriensis* is native but *P. pyrifolia* is alien. Thus, the hybridization would have occurred between wild trees or local landraces of the former and introduced plants of the latter. The time of hybridization is uncertain, but is perhaps not recent. Floristic studies show that trees referred to as *P. ussuriensis* with fairly large fruits occur in nature (Yü & Ku, 1974; Ku & Sponberg, 2003). They are perhaps hybrids, as found in Japan.

4. Future research possibilities

Our case of *Pyrus* suggests that whether a plant is native or alien could be inferred from molecular data. However, studies based on this strategy are rare. In Japan, Sasanuma et al. (2002) and Hori et al. (2006) reported the genetic uniformity of *Elymus humidus* and *Lycoris radiata*, respectively, both of which are prehistorically naturalized plants, in comparison with a certain level of genetic diversity among their wild relatives. European studies have focused on cultivated plants and their relatives. In addition to the abovementioned studies of *Daucus*, *Beta*, *Brassica*, and *Malus*, there are several studies of *Triticum* and *Aegilops* (e.g., Zaharieva & Monneveux, 2006; Arrigo et al., 2011). However, investigations in Europe would be difficult, because Europe is contiguous with the true native habitats of many archaeophytes, and archaeophytes could have been introduced much earlier (from the 6th millennium BCE) than in Japan. Nevertheless, as there are many more archaeophytes in Europe than in Japan and as many putative hybrids between native and alien plants have already been reported (Vila et al., 2000; Pyšek et al., 2002), many opportunities for research remain. For example, English elm (*Ulmus minor* var. *vulgaris*) was proved to be a 2000-year-old Roman clone (Gil et al., 2004).

The importance of research in East Asia and other parts of Eurasia is obvious (Castri, 1989). In research on alien plants, basic biological data such as floristic and ecological status and phytogeography are necessary. Research in historical documents and archaeological data is important for the assessment of archaeophytes. Fortunately, East Asian countries are rich in these resources. Traditional Eastern herbalism, which originated in China and was also developed in Japan and Korea, may provide plant records. Although research in the East still has many constraints compared with that in the West, our long historical cultural heritage offers one important advantage.

5. References

Aldasoro, J.J., Aedo, C., & Garmendia, F.M. (1996). The genus *Pyrus* L. (Rosaceae) in southwest Europe and North Africa, *Botanical Journal of the Linnean Society*, Vol. 121, No. 2, (June 1996), pp. 143–158, ISSN 1095-8339

Andersen, N.S., Poulsen, G., Andersen, B. A., Kiær, L.P., D'Hertefeldt, T., Wilkinson, M.J., & Jørgensen, R.B. (2009). Processes affecting genetic structure and conservation: a case study of wild and cultivated *Brassica rapa*. *Genetic Resources and Crop Evolution*, Vol. 56, No. 2, (March 2009), pp. 189–200, ISSN 0925-9864

Arnaud, J.F., Viard, F., Delescluse, M., & Cuguen, J. (2003). Evidence for gene flow via seed dispersal from crop to wild relatives in *Beta vulgaris* (Chenopodiaceae): consequences for the release of genetically modified crop species with weedy lineages. *Proceedings of the Royal Society of London, Series B*, Vol. 270, No. 1524 (August 2003), pp. 1565–1571, ISSN 0080-4649

Arrigo, N., Guadagnuolo, R., Lappe, S., Pasche, S., Parisod, C., & Felber, F. (2011). Gene flow between wheat and wild relatives: empirical evidence from *Aegilops geniculata, Ae. neglecta* and *Ae. triuncialis. Evolutionary Applications*, online, ISSN 1752-4571

Australian Weeds Committee (2011). Noxious weed list for Australian states and territories Ver.24.00, 15.09.2011, Available from
http://www.weeds.org.au/docs/weednet6.pdf

California Invasive Plant Council (2006). *California Invasive Plant Inventory*, Cal-IPC Publication, Retrieved from
http://www.cal-ipc.org/ip/inventory/pdf/Inventory2006.pdf

Castri, F. di (1989). History of Biological Invasions with Special Emphasis on the Old World, In: *Biological Invasions – A Global Perspective.* Drake, J.A., Mooney, H.A., Castri, F. di, Groves, R.H., Kruger, F.J., Rejmánek, M., & Williamson, M. (Eds.), ISBN 0471920851, John Wiley & Sons, New York, USA, pp. 1–30. Retrieved from
http://www.scopenvironment.org/downloadpubs/scope37/scope37-ch01.pdf

Cheffings, C. & Farrell, L. (Eds.) (2005). *The vascular plant Red Data List for Great Britain.* 15.09.2011, Available from http://jncc.defra.gov.uk/page-3354

Coart, E., Vekemans, X., Smulders, M.J., Wagner, I., Van Huylenbroeck, J., Van Bockstaele, E., & Roldan-Ruiz, I. (2003). Genetic variation in the endangered wild apple (*Malus sylvestris* (L.) Mill.) in Belgium as revealed by amplified fragment length polymorphism and microsatellite markers. *Molecular Ecology*, Vol. 12, No. 4, (April 2003), pp. 845–57, ISSN 0962-1083

Coart, E., Van Glabeke, S., De Loose, M., Larsen, A.S., & Roldán-Ruiz, I. (2006). Chloroplast diversity in the genus *Malus*: new insights into the relationship between the European wild apple (*Malus sylvestris* (L.) Mill.) and the domesticated apple (*Malus domestica* Borkh.). *Molecular Ecology*, Vol. 15, No. 8, (July 2006), pp. 2171–2182, ISSN 0962-1083

Culley, T.M. & Hardiman, N.A. (2009). The role of intraspecific hybridization in the evolution of invasiveness: a case study of the ornamental pea tree *Pyrus calleryana. Biological Invasions*, Vol. 11, No. 5, (May 2009), pp. 1107–1119, ISSN 1387-3547

Ellstrand, N.C. (2003). *Dangerous liaisons? When cultivated plants mate with their wild relatives,* Johns Hopkins University Press, ISBN 0-8018-74-5-X, Baltimore, MD, USA

European and Mediterranean Plant Protection Organization (2011). Invasive alien plants – EPPO lists and documentation, 15.09.2011, Available from
http://www.eppo.org/INVASIVE_PLANTS/ias_plants.htm

Fénart, S., Arnaud, J.F., De Cauwer, I., & Cuguen, J. (2008). Nuclear and cytoplasmic genetic diversity in weed beet and sugar beet accessions compared to wild relatives: new insights into the genetic relationships within the *Beta vulgaris* complex species. *Theoretical and Applied Genetics*, Vol. 116, No. 8, (May 2008), pp. 1063–1077, ISSN 0040-5752

Fitzpatrick, B.M., & Shaffe, H.B. (2007). Hybrid vigor between native and introduced salamanders raises new challenges for conservation. *Proceedings of the National Academy of Science of the United States of America*, Vol. 104, No. 40 (October 2007), pp. 15793–15798, ISSN 0027-8424

Food and Agriculture Organization of the United Nations [FAO] (2005). *Procedures for Weed Risk Assessment.* 15.09.2011, Available from
ftp://ftp.fao.org/docrep/fao/009/y5885e/y5885e00.pdf

Gil, L., Fuentes-Utrilla, P., Soto, Á., Cervera, M.T., & Collada, C. (2004). English elm is a 2,000-year-old Roman clone. *Nature* Vol. 431, No. 7012 (October 2004), pp. 1053, ISSN 0028-0836

Hardiman N.A., & Culley T.M. (2010). Reproductive success of cultivated *Pyrus calleryana* (Rosaceae) and establishment ability of invasive, hybrid progeny. *American Journal of Botany*, (October 2010) Vol. 97, No. 10, pp. 1698-1706, ISSN 0002-9122

Hawaiian Ecosystems at Risk project (2011). Hawaii's Most Invasive Horticultural Plants. In: *Hawaiian Ecosystems at Risk project*, 15.09.2011, Available from http://www.state.hi.us/dlnr/dofaw/hortweeds/specieslist.htm

Hori, T., Hayashi, A., Sasanuma, T., & Kurita, S. (2006). Genetic variations in the chloroplast genome and phylogenetic clustering of *Lycoris* species. *Genes & Genetic Systems*, Vol. 81, No. 4, (October 2006), pp. 243–253, ISSN 1341-7568

Iketani, H. & Ohashi, H. (2001). Rosaceae Subfam. II. Maloideae (=Pomoideae). In: *Flora of Japan IIb. Angiospermae Dicotyledoneae Archichlamydeae*, Iwatsuki, K., Boufford, D.E., Ohba, H. (Eds.), 111–124, Kodansha, ISBN 4-0615-4605-8, Tokyo, Japan

Iketani, H., & Ohashi, H. (2003). Taxonomy and distribution of Japanese populations of *Pyrus ussuriensis* Maxim. (Rosaceae). *Journal of Japanese Botany*, Vol. 78, No. 3, (June 2003), pp. 119–134, ISSN 0022-2062

Iketani, H., Yamamoto, T., Katayama, H., Uematsu, C., Mase, N. & Sato., Y. (2010). Introgression between native and prehistorically naturalized (archaeophytic) wild pear (*Pyrus* spp.) populations in Northern Tohoku, Northeast Japan. *Conservation Genetics*, Vol. 11, No. 1, (February 2010), pp. 115–126, ISSN 1566-0621

International Union for Conservation of Nature and Natural Resources [IUCN] (2000). *IUCN guidelines for the prevention of biodiversity loss caused by alien invasive species*. 15.09.2011, Available from http://www.issg.org/pdf/ guidelines_iucn.pdf

Iwatsuki, K., Yamazaki, T., Boufford, D.E. & Ohba, H. (Eds.). (1993–2011). *Flora of Japan Vol. 1–4*, Kodansha Scientific, Tokyo, Japan

Katayama, H. & Uematsu, C. (2006). Pear (*Pyrus* species) genetic resources in Iwate, Japan. *Genetics Resources and Crop Evolution*, Vol. 53, No. 3, (May 2006), pp. 483–498, ISSN 0925-9864

Keally, C.T. (2009). *Japanese Archaeology*, 15.09.2011, Available from http://www.t-net.ne.jp/~keally/index.htm

Kitamura, S. (1979). *Pyrus*, In: *Coloured Illustrations of Woody Plants of Japan Vol. 2*, Kitamura, S. & Murata, G. (Eds.), pp. 42–47, Hoikusha, Osaka, Japan (in Japanese)

Kobayashi, A. (1990). *Bunka to Kudamono (Culture and Fruits)*, Yokendo, ISBN 4-8425-9009-2, Tokyo, Japan (in Japanese)

Kovanda, M. (1965). On the genetic concepts in the Pomoideae. *Preslia*, Vol. 37, pp. 27–34, ISSN 0032-7786

Kowarik, I. (1995). Time lags in biological invasions with regard to the success and failure of alien species. In: *Plant Invasions: General Aspects and Special Problems*, Pyšek, P., Prach, K., Rejmánek, M., & Wade, M. (Eds.). pp. 15–38, SPB Academic Publishing, ISBN 90-5103-097-5, Amsterdam, the Netherlands

Kowarik, I. (2003). Human Agency in Biological Invasions: Secondary Releases Foster Naturalisation and population expansion of alien plant species. *Biological Invasions*, Vol. 5, No. 4, (December 2003), pp. 293–312, ISSN 1387-3547

Ku, T.C. & Sponberg, S.A. (2003). *Pyrus*. In: *Flora of China, Vol. 9, Pittosporaceae through Connaraceae*, Wu, Z.Y., Raven, P.H., & Hong, D.Y. (Eds.). pp. 173–179, Science Press, ISBN 7030111729, Beijing, China

Larsen, A.S., Asmussen, C.B., Coart, E., Olrik, D.C., & Kjær, E.D. (2006). Hybridization and genetic variation in Danish populations of European crab apple (*Malus sylvestris*). *Tree Genetics & Genomes*, Vol. 2, No. 2, (April 2006), pp. 86–97, ISSN 1614-2942

Maekawa, F. (1943). Prehistoric-naturalized plants to Japan proper. *Acta Phytotaxomica et Geobotanica*, Vol. 13, (November 1943), pp. 274–279, ISSN 0001-6799 (in Japanese)

Magnussen, L.S. & Hauser T.P. (2007). Hybrids between cultivated and wild carrots in natural populations in Denmark. *Heredity*, Vol. 99, No. 2, (August 2007), pp. 185–192, ISSN 0018-067X

Matsumoto, T., Takakura, K., & Nishida, T. (2010). Alien pollen grains interfere with the reproductive success of native congener. *Biological Invasions*, Vol. 12, No. 6, pp. 1617–1626, ISSN 1387-3547

Ministry of the Environment, Japan (2007). The Updated Japanese Red Lists on Mammals, Brackish-water/Freshwater Fishes, Insects, Shellfish, and Plants I and II. 15.09.2011, Available from http://www.env.go.jp/en/headline/headline.php?serial=503

Ministry of the Environment, New Zealand (2011). Alien plants which are considered weeds on conservation lands in New Zealand, 15.09.2011, Available from http://www.mfe.govt.nz/publications/ser/ser1997/html/tables/table9.4.html

Mittermeier, R.A; Gil, P.R., Hoffman, M., Pilgrim, J., Brooks, T., Mittermeier, C.G., Lamoreux, J., & da Fonseca, G.A.B. (Eds.). (2004). *Hotspots Revisited: Earth's Biologically Richest and Most Threatened Terrestrial Ecoregions*, University of Chicago Press, ISBN 978-9686-3977-72, Chicago, USA

Nakai, T. (1918). Notulae ad plantas japonicae et koreae XVI. *Botanical Magazine, Tokyo*, Vol. 32, No. 374, pp. 28–37, ISSN 0006-808X

Nentwig, W. (Ed.). (2003). *Biological Invasions*. Springer, ISBN 978-3-540-46919-6, Berlin, Germany

Ohashi, H. (1987). Floristic regions in the Tohoku District of Japan. *Journal of Japanese Botany*, Vol. 62, pp. 119–126, ISSN 0022-2062 (in Japanese)

Ohashi, H. (1993). Rosaceae. In: *Flora of Taiwan, 2nd Ed., Vol. 3*, Editorial Committee of the Flora of Taiwan (Eds.) pp. 69–157, Editorial Committee of the Flora of Taiwan, ISBN 957-9019-41-X, Taipei, Taiwan

Ohwi, J. (1965). *Flora of Japan*, Smithsonian Institute, Washington DC, USA

Potter D., Eriksson, T., Evans, R.C., Oh, S., Smedmark, J.E.E., Morgan, D.R., Kerr, M., Robertson, K.R., Arsenault, M., Dickinson T.A., & Campbell, C.S. (2007). Phylogeny and classification of Rosaceae. *Plant Systematics and Evolution*, Vol. 266, No. 1–2, pp. 5–43, ISSN: 0378-2697

Preston, C.D., Pearman, D.A., & Hall, A.R. (2004). Archaeophytes in Britain. *Botanical Journal of the Linnean Society*, Vol. 145, No. 3, (July 2004), pp. 257–294, ISSN 1095-8339

Pyšek, P. & Jarošík V. (2005). Residence time determines the distribution of alien plants. In: *Invasive plants: ecological and agricultural aspects*, Inderjit, B.P.C. (Ed.). pp. 77–96, Birkhäuser Verlag, Basel, Switzerland

Pyšek, P., Prach, K., Rejmánek, M., & Wade, M. (Eds.). (1995). *Plant Invasions: General Aspects and Special Problems*, SPB Academic Publishing, ISBN 90-5103-097-5, Amsterdam, the Netherlands

Pyšek, P., Sádlo, J., & Mandák, B. (2002). Catalogue of alien plants of the Czech Republic. *Preslia*, Vol. 74, No. 2, pp. 97–186, ISSN 0032-7786

Pyšek, P., Richardson, D.M., Rejmánek, M., Webster, G., Williamson, M., & Kirschner, J. (2004). Alien plants in checklists and floras: towards better communication between taxonomists and ecologists. *Taxon*, Vol. 53, No. 1, (February 2004), pp. 131–143, ISSN 0040-0262

Robertson, K.R., Phipps, J.B., Rohrer, J.R., & Smith, P.G. (1991). A synopsis of genera in Maloideae (Rosaceae). *Systematic Botany*, Vol. 16, No. 2, (April–June 1991), pp. 376–394, ISSN 0363-6445

Sasanuma, T., Endo, T.R., & Ban, T. (2002). Genetic diversity of three *Elymus* species indigenous to Japan and East Asia (*E. tsukushiensis*, *E. humidus* and *E. dahuricus*) detected by AFLP. *Genes & Genetic Systems*, Vol. 77, No. 6, pp. 429–438, ISSN 1341-7568

Shimizu, T. (Ed.). (2003). *Naturalized Plants of Japan*, Heibonsha, ISBN 4-582-53508-9, Tokyo, Japan (in Japanese)

Shōda, S. (2007). A Comment on the Yayoi Period Dating Controversy. *Bulletin of the Society for East Asian Archaeology*, Vol. 1, pp. 1–7, ISSN 1864-6026

Sukopp, H. & Sukopp, U. (1993). Ecological long-term effects of cultigens becoming feral and of naturalization of non-native species. *Cellular and Molecular Life Sciences*, Vol. 49, No. 3, (March 1993), pp. 210–218, ISSN 1420-682X

Takakura, K., Matsumoto, T., Nishida, T., & Nishida, S. (2011). Effective range of reproductive interference exerted by an alien dandelion, *Taraxacum officinale*, on a native congener. *Journal of Plant Research*, Vol. 124, No. 2, (March 2011), pp. 269–276, ISSN 0918-9440

Totman, C. (1987). *The Green Archipelago: Forestry in Pre-Industrial Japan*, University of California Press, ISBN 978-0520-0631-29, Berkeley, USA

Tutin, T.G., Heywood, V.H., Burges, N.A., Moore, D.M., Valentine, D.H., Walters, S.M., & Webb D.A. (Eds.) (1968). *Flora Europaea Vol. 2, Rosaceae to Umbelliferae*, Cambridge University Press, ISBN 052106662X, Cambridge, UK

Ueda, K. (1989). Phytogeography of Tôkai Hilly Land Element I. Definition. *Acta Phytotaxonomica and Geobotanica* Vol. 40, No. 1–4, (July 1989), pp. 190–202, ISSN 0001-6799 (in Japanese)

Vilà, M., Weber E., & D'Antonio, C.M. (2000). Conservation Implications of Invasion by Plant Hybridization. *Biological Invasions*, Vol. 2, No. 3 (September 2000), pp. 207–217, ISSN 1387-3547

Vincent, M.A (2005). On the Spread and Current Distribution of *Pyrus calleryana* in the United States. *Castanea*, Vol. 70, No. 1, (Mar 2005), pp. 20–31, ISSN 0008-7475

Willis, K.J. & Birks H.J.B. (2006). What is natural? The need for a longterm perspective in biodiversity conservation. *Science*, Vol. 314, No. 5803, (November 2006), pp. 1261–1265, ISSN 0036-8075

Wu, Z. & Raven, P.H. & Hong, D.Y. (Eds.) (2003). *Flora of China Vol. 9, Pittosporaceae through Connaraceae*, Science Press, ISBN 7030111729, Beijing, China

Yü, T.T. (1979). *Zhong guo guo shu fen lei xue (Classification of the fruit trees in China)*, Agriculture Press, Beijing (in Chinese)

Yü, T.T. & Ku, T.C. (1974). *Pyrus*. In: *Flora Reipublicae Popularis Sinicae, Vol. 36*, Yü, T.T. (Ed.), pp. 354–372, Science Press, Beijing, China (in Chinese)

Zaharieva, M. & Monneveux, P. (2006). Spontaneous hybridization between bread wheat (*Triticum aestivum* L.) and its wild relatives in Europe. *Crop Science*, Vol. 46, No. 2, pp. 512–527, ISSN 0011-183X

4

Amazonian Manatee Urinalysis: Conservation Applications

Tatyanna Mariúcha de Araújo Pantoja[1], Fernando César Weber Rosas[2],
Vera Maria Ferreira Da Silva[2] and Ângela Maria Fernandes Dos Santos[3]
[1]Museu Paraense Emílio Goeldi MPEG, Universidade Federal do Pará UFPA
[2]Departamento de Biologia Aquática e Limnologia, Divisão de Ictiologia e Mamíferos
Aquáticos, Instituto Nacional de Pesquisas da Amazônia INPA
[3]Instituto de Hematologia e Hemoterapia do Amazonas – HEMOAM
Brazil

1. Introduction

The Amazonian manatee (*Trichechus inunguis* Natterer, 1883) is an aquatic mammal (Family Trichechidae) that inhabits freshwater environments. It is endemic to the Amazon Basin, and occurs from Marajó Island (at the mouth of the Amazon River in Brazil) to the headwaters of the floodplain in Colombia, Peru and Ecuador.

Historically, the Amazonian manatee has been subjected to strong hunting pressure, and was a source of food not only for indigenous and fishery communities of the Amazonian region, but also a target of large-scale commercial fisheries throughout the 19th and early 20th centuries (Best, 1984; Rosas & Pimentel, 2001). In the year 1650 tons of meat and fat of these animals were sent to Europe (Best, 1982, 1984, Da Silva & Best, 1979; Junk and Da Silva, 1997). Later, between 1935 and 1954, its skin was used industrially. Due to durability, this material was used for the manufacture of pulleys, belts and hoses (Best, 1982, 1984), but with the advent of synthetic materials, its use in industry became less common (Rosas, 1991; Vianna et al., 2006

Today, despite being illegal, Amazonian manatees are still hunted (Rosas, 1994) since their flesh is still commonly consumed regionally. Manatee bones, skin and fat are also used for commercial purposes such as drug manufacturing. (M. Marmontel pers. comm., 2009). Due to the persistence of hunting, the animal has been listed as endangered species by the Instituto Brasileiro de Meio Ambiente e dos Recursos Naturais Renováveis - IBAMA Portaria nº 1.522/89 - and included in the International List of endangered animals of the International Union for Conservation of Nature - IUCN, as a vulnerable species (Ayres & Best, 1979, Costa et al. 2005; MMA, 2001, Rosas, 1991, Trujillo et al., 2006, 2008, Vianna et al., 2006). In Colombia, the species is also included in the Red Book of endangered species (Trujillo et al., 2006). Although protected by law in Brazil since the 1967 (Rosas, 1994), subsistence hunting, and to a lesser degree, commercial hunting – both of which still persist – have kept the species among those with "vulnerable" status (Hilton-Taylor, 2000).

In addition to hunting, other factors threaten the populations of the Amazonian manatee through occasional degradation and even loss of habitat for the species, leading to a decreased availability of habitat for performing key events in its cycle life (feeding, reproduction, etc.). Deforestation of riparian environments, water pollution (a serious threat to an herbivorous aquatic species, by damaging its habitat and a decrease of dietary plants), construction of hydroelectric plants, and accidental capture by fishing nets are also potential risks to this species (Rosas, 1991; Trujillo et al., 2006, 2008).

Urinalysis can be useful in determining the health status of captive animals. In this context, chemical, physical and sedimentological urinalysis could be a useful tool to monitor the health status of wild Amazonian manatees , increase our knowledge of its physiology, and provide a scientific foundation for future physiological studies of this species.

Due to the difficulty of obtaining biological samples, few data are available on the characterization, collection and production of urine in aquatic mammals in general, and very little is known about the composition of Amazonian manatee urine. The only available information on the urine of this species derives from analyses conducted in the Laboratory of Aquatic Mammals (LMA) of the Instituto Nacional de Pesquisas da Amazônia (INPA), Manaus, Amazon, Brazil. Some of these results were recently published (Pantoja et al., 2010), using urinalysis in order to establish normal ranges of urinary parameters to help monitor the health of this species in captivity. To accomplish this, the authors performed chemical urinalysis to obtain quantitative values of glucose, urea, creatinine, uric acid and amylase levels obtained using colorimetric spectrophotometry.

Other important information derived from analyses conducted over thirty years ago with seven captive specimens kept in the Laboratory of Aquatic Mammals (LMA) of the same institute. This information came from urine collection carried out by Robin Best with captive Amazonian manatees in 1978. These data were not published, but could be accessed from LMA's files, and were used as a comparison for the data obtained in this investigation, which presents the results from chemical, physical and sedimentological analysis of Amazonian manatees' urine in order to continue research useful for monitoring health in captivity.

We believe that any decisions about an animal's release or return to the wild should be made with caution and security. The safety of the individual animal as well as the continued well-being of the wild host population should be carefully considered. The information provided by this study can help support scientifically based reintroduction programs to re-establish or reinforce endangered native wild populations of Amazonian manatee. Urinalysis can be a useful tool since it can help assess the health status of animals proposed to be released or returned to the wild.

By establishing baseline data on physiological parameters using healthy captive animals, this study may also contribute to Amazonian manatee conservation since urine collection from wild manatees can help evaluate the condition of wild populations. Likewise, urinalysis could be used to monitor the health of other captive animals, such as those that may be reintroduced to former areas of habitat from which the species was removed by excessive hunting. Finally, this study extends our knowledge of the manatee's basic physiology, hopefully contributing to its effective management and conservation.

2. Material and methods

All animals studied were judged to be clinically healthy, based on their general appearance. Within a twelve-month period, 21 animals were sampled - nine females (F) and twelve males (M), classified into the following age classes: calves (0-2 years old/2 F and 2 M), juveniles (3-5 years old/4 F and 3 M), sub adults (6-9 years old/1 F and 4 M) and adults (over 10 years old/2 F and 3 M).

All juveniles, sub adults and adults were kept in three big pools (197 m³ each) and the calves were distributed in four smaller pools (6.4 m³ each). The animals were fed with grass (*Brachiaria mutica*), lettuce, cabbage and other vegetables at approximately 10% of their body weight per day. In addition, the calves were nursed by their mothers or fed with an artificial milk formula (Rodriguez et al., 1999).

Urine was collected once a month when the tanks were drained, by placing stainless steel containers under the genital slit of females until their urination (Fig. 1a). Males were turned on their side and abdominal massages were applied to stimulate micturition (Bossart et al., 2001) (Fig. 1b). When this latter procedure did not work, the same method employed to females was applied to males.

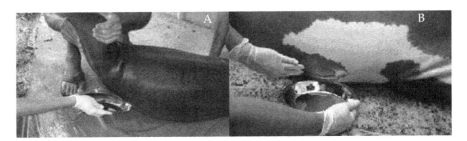

Fig. 1. A) Stainless steel containers being placed under the genital slit of female Amazonian manatee until their urination (Photo: Mattos, G. E.); B) Urine collection in male Amazonian manatee, turned on its side, with belly being massaged to stimulate urination. (Photo: D'Affonseca Neto, J. A.).

Chemical urinalysis was conducted by using dip strips (URISCAN™) just after urine collection of each sample by soaking the strips in fresh urine and making a comparison of the resulting color of the strip with the standardized color chart provided with the kit. Urinary parameters measured by dip strips were: protein, ketones, urobilinogen, bilirubin, pH, blood cells (erythrocytes), leukocytes, nitrite, specific gravity (actually a physical parameter), and glucose, which was also analyzed quantitatively by colorimetric spectrophotometry chemistry analyzer "Dimension AR" (Dade Behring) after centrifugation of samples to 300 RPM. All these compounds are physiologically related to homeostasis equilibrium, whose monitoring can be a useful tool to management actions (e.g. environment protection programs that aim to reintroduce these mammals to their natural environment). Urinary compound levels are indicative of the health status of manatees and other aquatic mammals (Gürtler et al., 1987; M. S. Matos & P. F. Matos, 1995; Kantek Garcia-Navarro, 1996).

The physical examination consisted in the observation of color, appearance and density of urine samples. The first two were made with the naked eye, in conjunction with sediment analysis, always using good light for easy viewing and characterization (Strasinger, 1996). The terminology used for recording urine color was: colorless, yellow, lemon yellow and reddish. Appearance was recorded as: clear, semi-cloudy and cloudy. Although the specific gravity can be considered a physical parameter, this was considered along with other chemical parameters following the method described for the reactive strips.

For sediment urinalysis, samples were centrifuged and the remaining solid was placed between slide and coverslip for examination under a light microscope (10x and 40x magnification). Elements present in the urinary sediment were counted per visual field (10X magnification). Leukocytes and erythrocytes were counted in five visual fields and then averaged between them for each urine sample. Epithelial cells and crystals were recorded as absent, rare (only a few elements in the visual field), frequent (half the visual field containing these elements) or numerous (elements filling the entire visual field). Bacterial abundance was recorded as "absent", "low" (few bacteria per visual field), moderate (half the visual field containing bacteria) and "high" (bacteria fill the entire visual field). For hyphae, these were recorded as present or absent. Little light was used to facilitate the search of urinary sediment elements (Bossart et al. 2001; Kantek Garcia-Navarro, 1996).

Qualitative results obtained by reactive strips (protein, ketones, urobilinogen, bilirubin, pH, blood cells (RBCs), leukocytes, nitrite, specific gravity and glucose) and data obtained from analysis of color, appearance and sediment were grouped in tables, according to sex and age. Some elements of the sediment were photographed through a camera attached to the microscope and these results (boards with such elements) will be presented throughout the chapter.

3. Results and discussion

A total of 188 tests were performed, 108 of them with urine samples of males and 80 with urine samples from females[1].

3.1 Chemical urinalysis

3.1.1 Protein

Protein levels indicated by reactive strips were minimal (<10 mg/dL) and constant throughout the experiment, for both males and females, and for different age groups.

Normally, proteins do not cross the glomerular membrane. When this occurs, it is in a small quantity which is then reabsorbed in the tubules. If some protein is not reabsorbed, a little can be present in urine but in such small quantities that can remain undetectable, or at most, appears as "traces". This can occur in more concentrated urine, like the first morning urination or that produced after muscular effort. Proteinuria (protein in urine) may have a renal or a post-renal origin, and the former may be physiological or caused by a glomerular

---

[1]In the first month of collection (June) it was not possible to collect urine from a juvenile female (Adana).

lesion (Gürtler et al., 1987) characterized by the presence of cylinders in the sediment. Proteinuria without casts (cylinders in the urine), suggests that its origin is post-renal, or passing, without further associated injury (Kantek Garcia-Navarro, 1996).

Low levels of protein (1 g/L (100mg/dL)) were detected in *Halichoerus grypus* (gray seal) urine by Schweigert (1993). Although low, these values were higher than those detected in this study in Amazonian manatees, possibly due to the difference in protein consumption, which is lower in *T. inunguis,* because it is an herbivorous animal.

Despite the small sample size, the constant and negative results observed over the nine months of the experiment (<10 mg/dL) ruled out the possibility of the occurrence of pathological or physiological proteinuria in manatees sampled and these results were considered as normal for the qualitative detection of proteins by reactive strips for the species.

3.1.2 Ketones

The measured levels of ketones in urine samples were also constant for 100% of the samples, both for males and females in different age groups, and always corresponded to the minimum detectable by reactive strips (<5 mg/dL).

When the diet is low in carbohydrates, the body uses its fat reserves for energy. The metabolism of fatty acids results in the formation of so-called ketone bodies (ketones), which when excessive can be observed in urine (ketonuria). Such an occurrence may be indicative of: 1) diabetes mellitus (usually accompanied by hyperglycemia and glucosuria), 2) starvation (as ketone bodies are just from the metabolism of body fat stores in case of famine), 3) intake grazing of poor quality, i.e., containing a few digestible carbohydrates, in the case of cattle, 4) prolonged vomiting or diarrhea (which cause ketonuria similar to that caused by starvation), or 5) other causes such as acute febrile diseases and toxic states (especially when accompanied by vomiting or diarrhea), endocrine disorders such as hyperactivity of the anterior pituitary and adrenal cortex, acute or chronic liver disease (Kantek Garcia-Navarro, 1996), and ketosis in ruminants (Gürtler et al ., 1987).

A non-detection of ketone bodies by qualitative analysis using reactive strips in urine from the gray seal was reported by Schweigert (1993). Similar results were obtained in this study with Amazonian manatees, and the constant qualitative results of ketone bodies (<5 mg/dL, negative value) suggests, in addition to discarding the hypothesis of occurrence of any of the pathologies listed above, that this should be the normal value expected for *T. inunguis* urine.

3.1.3 Urobilinogen

Urobilinogen values detected in urine samples from males and females did not vary throughout the experiment and showed minimum value measured by reactive strips (0.1 mg/dL).

Urobilinogen is the reduced form of conjugated bilirubin. The elimination of a part of this component in urine is normal, since it is the main urinary pigment that gives yellow coloration to the urine. The main causes of urobilinogenuria (increased urine urobilinogen) are: 1) liver disease (in which urobilinogen gets to be detected in dilutions up to 1:40) or 2) hemolytic jaundice. The decrease in urinary urobilinogen, in turn, may be due to: 1) total or partial obstruction of the bile ducts, 2) diarrhea (which may decrease the intestinal absorption of urobilinogen, causing a slowdown in its elimination), 3) action of certain antibiotics that cause inhibition of normal intestinal flora that are responsible for its production, or 4) chronic nephritis (that is accompanied by polyuria, which causes dilution of urobilinogen in urine) (Kantek Garcia-Navarro, 1996).

The constant results (0.1 mg/dL, negative values) ruled out urobilinogenuria and suggest that these values can be considered as normal for qualitative research by reactive strips of urobilinogen in the urine of Amazonian manatee. The possibility of a reduced elimination of urobilinogen in urine can be ruled out, since, in addition to the uniformity of the results, samples showed minimal yellowing.

3.1.4 Bilirubin

Except for an examination of a juvenile male (Mapixari) in July and two other examinations performed on two subadult males (Erê and Guarany) in November, which resulted in 0.5 mg/dL, bilirubin values obtained for all other males and all females were the least detectable by reactive strips (<0.5 mg/dL).

Bilirubin is a pigment derived from hemoglobin resulting from the degradation of erythrocytes by the macrophage-monocytic phagocytic system. When released by macrophages, bilirubin binds to albumin and thus does not appear in urine as the latter does not cross the glomerular barrier. Bilirubin in the liver turns off the albumin and conjugates with glucuronic acid to form conjugated bilirubin, which in turn is excreted in small amounts but not in all animals (only 20% of dogs, 5% of cats and 25% of cattle). An increase in bilirubin in urine (bilirubinuria) may be due to: 1) obstruction of the bile ducts, causing bile reflux (conjugated bilirubin) into the circulation, causing bilirubinuria, 2) liver disease (such as infectious canine hepatitis, leptospirosis, liver cirrhosis, various cancers of the liver, liver toxicosis and other) 3) hemolytic jaundice, or 4) intestinal obstruction (since there is decreased elimination of bilirubin through the gut, thereby increasing the rate of plasma conjugated bilirubin , which happens to be excreted in the urine) (M. S. Matos & P. F. Matos, 1995; Kantek Garcia-Navarro, 1996).

Negative results (<0.5 mg/dL) observed in most quality tests by reactive strips in both males and females of *T. inunguis,* and the results of sub-adult males (Erê and Guarany) and of a juvenile male (Mapixari) (which reached 0.5 mg/dL, still considered "traces") did not point to any of the conditions listed above. Low values of bilirubin, according to M. S. Matos & P. F. Matos (1995), can usually be found in 25% of healthy cattle, corroborating that the presence of bilirubin as "traces" does not confirm the occurrence of diseases. The absence of bilirubin was also reported in earlier studies performed in *T. inunguis* (Best, unpublished data[2]); in only one male (Xingu, 1/19/1978) were "light strokes" of bile pigments were

---

[2]This information came from urine collection carried out by Robin Best (in memorian) in captive Amazonian manatees in 1978. These data can be found in LMA's files, and it was possible to access them with the permission of the LMA directorships: Dr. Fernando César Weber Rosas and Dr. Vera Maria Ferreira da Silva.

detected. Therefore, negative values of bilirubin were regarded as the normal condition expected for *T. inunguis* urine. The "traces" of this pigment detected by reactive strips does not consist in a diagnosis of any of the physiological dysfunctions mentioned above.

3.1.5 pH

The minimum pH value measured by reactive strips in the urine of the Amazonian manatee was 5.0 and the maximum was 9.0. The pH 8.0 was the most observed in the samples (n=132), differing only in female infants, whose pH in most samples (n=4) was 6.0.

Urine pH is related to the maintenance of acid/base equilibrium maintained by the renal elimination of nonvolatile acids and alkalis. It is determined by metabolism and diet, being more acid in animals with high protein diets (meat eaters), and more basic in animals with diets high in carbohydrates (herbivores) (Gürtler et al. 1987; Kantek Garcia-Navarro, 1996).

With respect to organic metabolism, the pH of the urine normally accompanies the body pH, except in cases of aciduria (urine with an acid pH) found in the alkalosis existing in severe hypochloremia accompanied by vomiting (paradoxical aciduria) (Kantek Garcia-Navarro, 1996). Table 1 below shows the reference values of urinary pH of some domestic and laboratory animals, compared with Amazonian manatee detected values.

Animal group	pH	Reference
Cattle[H]	7.4 to 8.4	
Canidae[C]	5.5 to 7.5	
Felidae[C]	6.0 to 7.0	
Equine[H]	7.0 to 8.0	Benjamin (1976)
Sheep and goat[H]	7.0 to 8.0	
Pig[O]	5.5 to 8.5	
Human[O]	4.8 to 7.5	
Amazonian manatee[H]	6.0 to 9.0	Best (unpublished data)
Amazonian manatee[H]	5.0 to 9.0	This study

C=carnivorous; H=herbivorous, O=omnivorous

Table 1. Urinary pH values of some domestic and laboratory animals, humans, and the Amazonian manatee.

Urine acidification may be due to: 1) starvation (when the body uses its own plasma proteins resulting in an acidic urine), 2) respiratory or metabolic acidosis (the former occurs by the accumulation of CO_2 in the lungs in patients with emphysema, pulmonary fibrosis and cardiopulmonary failure, and the latter occurs when the body loses bicarbonate or accumulate acids, a condition observed in diabetes mellitus and chronic renal failure with uremia, i.e., a large quantity of urea in the blood), 3) rapid action drug therapy, such as methionine or sodium chloride, calcium, or ammonia, 4) increased protein catabolism, as with fever or great muscle effort, 5) rapid absorption of cavity fluids (which have high protein value); 6) paradoxical aciduria (mentioned above), or 7) diseases that cause tissue disintegrations, such as diabetes, Fanconi syndrome and proximal tubular acidosis (Kantek Garcia-Navarro, 1996).

On the other hand, urine alkalinization may be due to: 1) time taken to perform the exam (since there may be bacterial growth transforming urea to ammonia; however, in the present study this possibility was excluded by carrying out the pH determination immediately after urine collection) (Bossart et al., 2001), 2) cystitis, especially when accompanied by urinary retention, and subsequent bacterial action on urea, as above, 3) treatment with salts of alkaline reaction, such as baking soda, or sodium lactate, citrate or sodium or potassium acetate, potassium nitrate , acetazolamide and amphotericin B or 4) metabolic alkalosis (due to accumulation of bicarbonate), or respiratory failure (when there is increased ventilation resulting from CO_2 elimination at a greater rate than its production, a situation that can occur in acute cardiorespiratory disease accompanied by hypoxia) (Kantek Garcia-Navarro, 1996).

The range of pH values recorded in this study (5.0 to 9.0) was similar to that observed originally for *T. inunguis*, which was 6.0 to 9.0 (Best, unpublished data). In most of our qualitative analysis by reactive strips, detected pH was alkaline (pH=8.0) for both males (n=85) and for females (n=47). This finding corresponds to that expected for herbivore urine (Kantek Garcia Navarro, 1996; M. S. Matos & P. F. Matos, 1995), agreeing with the results found by Best (unpublished data). Manire et al. (2003) observed in two males (11 and 14 years) of *Trichechus manatus latirostris* an average pH=8.0 ±0.31 (n = 15) and 7.99 ±0.30 (n=13) – these values are very similar to those observed in most samples analyzed in this study. However, Bossart et al. (2001) had reported more acidic values in *T. manatus latirostris* (6.0 to 7.5).

More acidic pH values (between 5.0 and 6.0) were recorded in gray seals by Schweigert (1993), as well as for the southern sea lion (*Otaria flavescens*), whose pH values of 31 samples ranged from 5.69 to 7.0 (mean 6.25 ±0.26) (Le Bas, 2003). pH values around 6.0 were also observed in the urine of three freshwater dolphins *Inia geoffrensis, Platanista indi* and *P. gangetica* (De Monte & Pilleri, 1972), which have a high protein diet due to feeding on fish. In this study, the registration of more acidic pH (5.0 and 6.0) was not attributed to: 1) starvation, since none of the animals stopped eating at this point, 2) the action of acid-acting drugs, which were not administered to the manatees in the period of the experiment, or 3) any of the diseases listed above as possible causes of urinary acidification, since the animals were apparently healthy throughout the experiment. However, the following factors: 1) increased protein catabolism, caused by intense muscular effort (possibly caused when the animal was restrained for urine collection) and 2) rapid absorption of cavity fluids could be considered as well as normal individual variations. However, these factors need further studies to confirm occurrence in manatees as the cause of urinary acidification.

The small sample size did not allow a more robust statistical check on the changes in pH observed, and although the value 8.0 was the most frequent, there is a need for more conclusive studies that would allow the establishment of a normal value expected by qualitative assessment of this parameter by reactive strips.

3.1.6 Blood cells (erythrocytes – red blood cells RBC's)

Blood cells (RBCs) detection by reactive strips pointed to following values: <10, 10 and 50 RBC/mL. The value of 10 RBC/mL was observed in 86.7% of the tests (n=163) of both males and females of different age classes.

When there is blood in urine, it has reddish-brown color. This occurrence can sometimes manifest as haematuria, sometimes as hemoglobinuria. The difference will be made by the presence of erythrocytes (RBCs) in the sediment and by the appearance of urine. Haematuria is confirmed when red blood cells are present in whole urine, appearing cloudy. It follows from hemorrhage or renal urogenital tract, glomerulonephritis, vasculitis, or renal infarction (when red blood cells pass into the tubules). Hemoglobinuria, in turn, is the presence in solution of hemoglobin in the urine, cloudy and semi-looking brown or red (due to the fact that hemoglobin is free in plasma) (Kantek Garcia-Navarro, 1996).

According to its origin, hemoglobinuria can be true or false. In the true form, it appears in the urine half blurred due to the presence of free hemoglobin in plasma (hemoglobinemia). This hemoglobin passes through the glomerular barrier and, owing to its great quantity, is not fully reabsorbed and may be preset in urine suggesting intravascular hemolysis, which occurs in hemolytic anemia with intravascular hemolysis and hemoglobinemia. The latter include babebioses and hemolytic disease in newborn individuals. The hemoglobinemia with intravascular hemolysis and hemoglobinuria may be due to either 1) *Clostridium haemolyticum* or *C. perfringens* infections, 2) postpartum hemoglobinuria (M. S. Matos & P. F. Matos, 1995), 3) ingestion of certain toxic plants, 4) severe burns with destruction of large amounts of tissue, 5) the action of toxic agents (e.g. copper , mercury and sulfa drugs), or 6) the action of hemolysins produced by *Leptospira pomona* (Kantek Garcia-Navarro, 1996). Even if the latter could be considered in the diagnosis of *T. inunguis* since Marvulo et al. (2003) detected antibodies against *Leptospira sp.* in blood samples from two Amazonian manatees in captivity, suggesting that these animals had had contact with leptospires. There is a need for further studies to elucidate this finding.

Regarding the false presence of hemoglobin, this is a consequence of the breakdown of red blood cells present in very dilute or alkaline urine, a fact that usually occurs *in vitro*, and may also occur within the urinary bladder. In fact, it's a haematuria masquerading as hemoglobinuria. To confirm it is necessary to search erythrocytes in whole sediment, since such hemolysis, not rare, is partial (Kantek Garcia-Navarro, 1996).

The strips are highly sensitive to reactive detection of hemoglobin, but have low sensitivity to intact erythrocytes, requiring microscopic examination for their confirmation ("URISCAN™ Urine Strip" bull). The qualitative results obtained by reactive strips were disregarded, since in the sediment survey we observed different amounts of red blood cells detected by reactive strips. Therefore, in the present study, detection of red blood cells during microscopic analysis of the sediment was considered as the most efficient. Additionally, the "URISCAN™ Urine Strip" bull alert to the fact that reaction to the test strip may vary from one patient to another, and that even in case of detection of intact red blood cells or hemoglobin, its necessary to analyze the sediment for confirmation.

3.1.7 Leukocytes

The amount of leukocytes in the samples did not exceed 25 WBC/mL, and, for 90.4% of them (n=170) less than 25 WBC/mL were detected for both male and females of different age groups.

Leukocytes are white blood cells that may appear in urine as small cells, with round and granular cytoplasm; 7 per field is the amount considered normal for leukocytes in urine (Kantek Garcia-Navarro, 1996). In case of a reaction to an infection, it appears in its degenerate form (pus cells), and when present in large numbers may be indicative of disease (Gürtler et al., 1987) as inflammation of the renal system, prostate, uterus or vagina (M. S. Matos & P. F. Matos, 1995), or simply be due to fever or strenuous exercise (in this case appears only temporarily) (Kantek Garcia-Navarro, 1996). The quality tests for reactive strips for leukocytes detection in samples proved to be ineffective, as the results never coincided with those observed by microscopic analysis of the sediment, which was considered the most accurate research of leukocytes in samples.

3.1.8 Nitrite

The detection of nitrite by reactive strips in males was variable throughout the experiment. Despite the absence of nitrite in most samples of males (n=73), in a few months nitrite was detected in at least half of them. The survey of nitrite in the samples of females, in turn, followed a pattern throughout the experiment, indicating the absence in the urine analyzed, except the samples for the months of July and October, in an adult female (Tukano).

Nitrites can be produced by bacteria present in urine, so the presence of nitrite in urine may indicate urinary tract infection. Apparently, the test has not had the same use in animals and in humans, although a correlation was recently demonstrated between the presence of nitrites in urine and urinary tract infections in pigs (Kantek Garcia-Navarro, 1996).

Because the results for nitrite in urine is directly related to the presence of bacteria in the urine, it will be discussed later, along with discussion of elements of urinary sediment. It is noteworthy that the results of the presence or absence of nitrite in the urine, detected by reactive strips, also did not coincide with the appearance of bacteria in urine during microscopic analysis of the sediment, a method that, as in the case of leukocytes, was considered more accurate for confirming the presence of nitrite in the samples.

3.1.9 Specific gravity

Values observed in urine specific gravity of *T. inunguis* ranged from 1.000 to 1.015. The value 1.005 was observed in most samples (n=153). A single urine sample had density 1.015 (a female juvenile (Adana) in October).

The specific gravity of urine determines the ability of renal reabsorption. It defines the urine specific gravity compared to the density of the same volume of distilled water at the same temperature. Since urine is actually water that contains dissolved chemicals, urine specific gravity is nothing more than a measure of the density of these chemicals dissolved in the sample (Strasinger, 1996). Therefore, since the density measures the degree of solutes present in the sample, it evaluates renal ability to concentrate urine. The higher the density value, the more concentrated (hypertonic) is the urine, i.e., an increase in density indicates a decrease in glomerular filtration and/or increased water reabsorption, facts that lead to a large reduction in urine flow (olyguria) (Kantek Garcia-Navarro, 1996).

Decreased urine specific gravity typically accompanies polyuria (except in diabetes mellitus, in which polyuria is associated with high density), and may be due to: 1) uremia (clinical condition in which there is an increased rate of urea in the blood, by renal (primary) or systemic (secondary) cause, and in aggravated cases, density may be extremely low due to the inability of the kidney to dilute the urine), 2) diabetes insipidus, 3) uterine empyema (pyometra) which produces excessive thirst (polydipsia) and large increase in urine flow (polyuria) in female domestic dogs and 4) corticosteroids therapy, parenteral fluids or diuretics, usually accompanied by polyuria; 5) isosthenuria (urine with the same density of filtered plasma before passing through the glomerular meshwork) resulting from renal failure in dilute or concentrate urine; 6) chronic interstitial nephritis (due to kidney inability to concentrate urine), or 7) simple excessive water consumption (M. S. Matos & P. F. Matos, 1995; Kantek Garcia-Navarro, 1996).

Related to the increase in urine specific gravity, usually associated with olyguria (except diabetes mellitus), this can be seen in cases of: 1) acute interstitial nephritis (with density values between 1.030 and 1.060, due to the initial phase of the disease in which there is inability of renal elimination of water), 2) general acute nephritis (where there is reduced glomerular filtration rate (producing a more concentrated urine), 3) diabetes mellitus and primary renal glucosuria (an increased density accompanied by polyuria, as that glucose "load" a greater amount of water), 4) dehydration, sometimes due to vomiting, diarrhea, or excessive sweating as they reduce the amount of water available in the kidney, thus producing a more concentrated urine (values greater than 1.050 in dogs and cats up to 1.060, suggest severe dehydration), 5) fever (since body attempts to retain water, thus producing a more concentrated urine), 6) edema (due to a circulatory dysfunction, caused by excessive fluid retention in the body, resulting in olyguria with high density), 7) shock, hypotension which causes a sudden fall in renal perfusion, producing olyguria, which can reach the cessation of urine flow (anuria) (Kantek Garcia-Navarro, 1996); 8) cystitis, or 9) debilitating illnesses tissue destruction (M. S. Matos & P. F. Matos, 1995).

Although little is known about the maintenance of water in manatees, Maluf (1989) analyzing the anatomy of nine manatee kidneys (*Trichechus manatus*), suggested that they have an increased ability to concentrate urine. Since *Trichechus manatus* inhabits both freshwater and marine, it can be considered an interesting model for osmoregulation study in Sirenians. Drawing on this, Ortiz et al. (2001) conducted studies with that species, and concluded that manatees are excellent osmoregulators, regardless of the medium in which they live. In *T. inunguis,* the results of our analysis pointed to a little concentrated urine, tending to low density, which in freshwater animals, serve to expel the excess water, while solutes are retained (Schmidt-Nielsen, 2003). Low specific gravity was also reported by Manire et al. (2003) for two males of *T. manatus latirostris* (1.008 ±0.002, n=15 and 1.010 ±0.004, n=13) that did not respond significantly between the manipulations made in the experiment (reduction in the amount of food, change of diet cabbage, apples, carrots and beans to "seagrass" and change from freshwater to saltwater), indicating that manatees probably would not concentrate their urine to conserve fluid, even when in hemoconcentration (dehydration). The amplitude of density urinary density values measured by Best (unpublished data) was between 1.000 and 1.013, very similar to that observed in our study (1.000 to 1.015). Higher levels of density were observed by Schweigert

(1993) for gray seal, whose values ranged between 1.033 and 1.052, given the exceptional ability of marine mammals' reniculate kidneys to concentrate urine.

Although 1.005 has been the value observed in most samples (n=153), both in males and females, variations that occurred in some tests were considered acceptable, since they did not exceed 1.015 (maximum value observed in the samples analyzed in this study), which would be a much more concentrated urine than those previously reported for *Trichechus*. However, further studies are needed to establish a normal value expected from qualitative assessment of reactive strips for urine density in *T. inunguis*.

3.1.10 Glucose

In our previous research (Pantoja et al, 2010), we measured quantitative levels of glucose by a chemical analyzer, and found no statistically significant difference between sexes and age classes, leading to the establishment of a normal range (3.0 to 3.6 mg/dL) for this parameter. Since the present examination of glucose levels by reactive strips in both males and females, regardless of age group, was the minimum value detectable by the strips (<100 mg/dL), we should take this result as a confirmation of the adequacy of reactive strips as a rapid, cheap and efficient method to access this condition in manatee urine samples.

When glucose is present in normal amounts in the blood, it is absent in the urine, since it is completely reabsorbed in the proximal renal tubules, appearing only as traces in urine (Gürtler et al. 1987; M. S. Matos & P. F. Matos, 1995; Kantek Garcia- Navarro, 1996). Its appearance in urine occurs when the amount of glucose in the glomerular filtrate exceeds the capacity of the tubule reabsorption (when glucose is increased in the blood due to diabetes mellitus), or when there is insufficient tubular reabsorption. The occurrence of glucosuria (increased glucose in the urine), should be confirmed by measuring the amount of glucose in the blood of the animal during fasting (M. S. Matos & P. F. Matos, 1995; Kantek Garcia-Navarro, 1996).

Glucosuria may have physiological or pathological origin. The physiological glucosuria is usually transient and can result from ingestion of large amounts of carbohydrates. Emotional glucosuria may occur during animal restraint during urine collection, since a sudden release of adrenaline occurs. According Kantek Garcia-Navarro (1996) there are cases of glucosuria in domestic cats under physical stress, or severe bleeding in the bladder. When pathological, glucosuria is evident both in animals fasting or at rest and can indicate the following conditions: 1) diabetes mellitus, 2) acute pancreatic necrosis (when insulin production drops, thereby determining the subsequent hyperglycemia and glucosuria), 3) hyperthyroidism (and quick absorption of carbohydrates in the gastrointestinal tract), 4) acute renal failure, in which there is deficiency in the tubular reabsorption of glucose, or 5) chronic liver disease, in which there is an inability to regulate the liver glycogen stores (Kantek Garcia-Navarro, 1996). In the case of diabetes mellitus, glucose concentration varies according to the severity of the disease, which occurs mostly in older dogs and cats (Gürtler et al., 1987).

The consistently negative results of glucose (<100 mg/dL) measured qualitatively by reactive strips suggests that the expected levels of glucose in *T. inunguis* should be low. Quantitative analysis by means of chemical analysis "Dimension AR" ("Dade Behring") resulted in low levels of glucose, with an amplitude in males 0 to 10 mg/dL and in females 0 to 13 mg/dL. These values were considered very low, corroborating the results obtained by reactive strips. These low levels obtained by both methods ruled out the occurrence of glucosuria, and therefore the diseases related to it, and suggest the applicability of reactive strips in routine examinations in Amazonian manatees in captivity because of the ease, convenience and low cost of this method.

The failure to detect glucose was also reported in gray seals by Schweigert (1993). De Monte & Pilleri (1972) found traces of glucose in an individual of *Platanista minor* (Indus river dolphin), claiming that this was the first detection of this substance in cetaceans by these authors. The detection of high levels of glucose (200 mg/dL) was also reported by Ridgway et al. (1968) in one specimen of bottle-nosed dolphin *Tursiops truncatus* in captivity without clinical symptoms of diabetes. The data from urinalysis of manatees in the Amazon conducted by Best (unpublished data) did not find glucose in any examinations (n =13) of individuals of both sexes. Although not quantitative, and considering the low variation in food that these animals were given at that time (*Cabomba sp.* and *Brachiaria mutica*), compared with the more diverse diet currently offered to manatees in captivity, these data are consistent with the low glucose values observed in *T. inunguis* urine in this study.

3.2 Physical and sedimentological urinalysis

3.2.1 Urine color

Urine samples showed the following colors: colorless, yellow, lemon yellow and red. Most, both in males (n=73), and females (n=48) were characterized as "yellow". In female infants, half of the samples were yellow (n=5) and in subadult females this coloration appeared in four samples. The red coloration was seen in a male infant (Tuã) in January and February, in a subadult male (Guarany) in November, and in a subadult female (Cunhataí) in January.

The change in urine color is due to the amount of urinary pigments (urochrome and uroerythrin) (M. S. Matos & P. F. Matos, 1995; Kantek Garcia-Navarro, 1996; Strasinger, 1996), and to the elimination of drugs or metabolic products of organic dyeing properties, which may occur due to ingestion of certain foods, medications, or result from various physiological or pathological states (M. S. Matos & P. F. Matos, 1995; Kantek Garcia-Navarro, 1996).

In domestic animals, urine color varies from light yellow to dark brown. While the urine of omnivores and carnivores is usually pale yellow, the herbivores' ranges from light yellow to dark brown. The exact color of the urine depends on exogenous dyes (via food) and endogenous dyes (from hemoglobin and protein metabolism). The urine after voiding tends to be clear, however, in herbivores, darkening can occur after urine collection due to the presence of oxygen or through bacteria putrefaction (Gürtler et al., 1987).

De Monte & Pilleri (1972) reported a color ranging from light yellow to lemon yellow or olive for the urine of *I. geoffrensis*, *P. indi* and *P. gangetica*. The amber color of the urine was observed in other cetaceans and pinnipeds which, by feeding on fish, have a diet rich in animal protein, resulting in this color (Medway & Geraci, 1986, as cited in Bossart et al., 2001). Amber color was not observed in any sample of urine from Amazonian manatees.

However, according to Bossart et al. (2001) this is the color normally found in the urine of *T. manatus latirostris*.

The color "yellow", in polyuria occurs when urine is too diluted and the specific gravity is consequently low (except in the occurrence of diabetes mellitus, in which case urine is colorless, but with high specific gravity) (Kantek Garcia-Navarro, 1996). By tracking the urinary staining presented in the course of the experiment, it was observed that the yellow staining was more frequent in *T. inunguis* urine. The absence of color (urine "colorless") observed in some samples also appears to be a normal finding for this species, which presents urine with low specific gravity (more diluted). The color "red" may be due to haematuria or hemoglobinuria, or ingestion of sugar beet (Kantek Garcia-Navarro, 1996, Strasinger, 1996). In this study, urine samples that showed this color were certainly due to the ingestion of sugar beets, because sediment red blood cells were not detected in any of these samples by microscope examination. Moreover, the presence of sugar beet in the diet of manatees whose urine showed that coloration reinforces the assumption that this was the cause of staining observed. The staining results recorded during the experiment may be used in future comparative urinalysis in Amazonian manatees.

3.2.2 Appearance

Urine samples of Amazonian manatees showed the following appearance: clear, semi-cloudy and cloudy. In males, only two samples: in one juvenile male (Tapajós) in December and in a subadult male (Guarany) in September, showed cloudness. Most samples of females (n=63) presented semi cloudy aspect, however, some samples were clear or cloudy.

The aspect corresponds to a general term that refers to the transparency of the urine sample (Strasinger, 1996). Urine of any species may become cloudy if left to stand for some time due to precipitation of salts that may be present. Other causes of cloudy urine is the presence in large amounts of epithelial cells, erythrocytes, leukocytes, bacteria, mucus and crystals, from either urinary organs, sometimes of the genitals (Kantek Garcia-Navarro, 1996), and also the presence of lipids, semen, lymph, yeast, fecal matter or even external contamination (Strasinger, 1996). Appearance determination needs to be confirmed by the analysis of the sediment depending on the occurrence of each of these cellular elements.

Normal urine is usually transparent just after being excreted; however, it may acquire certain opacity generated by the precipitation of amorphous phosphates and carbonates in the form of white mist. The opacity caused by the presence of the above elements, especially epithelial cells in females, does not necessarily mean pathogenesis (Strasinger, 1996). The registration of some samples in this study classified as semi-turbid was due to the presence of such elements. The presence of large quantities of white blood cells, red blood cells or bacteria, on the other hand, may indeed be evidence of pathogenicity and the fact that the sample recently eliminated was blurred, may be a cause for concern (Strasinger, 1996).

None of the samples that presented high turbidity immediately after collection showed high amounts of substances that could indicate a serious illness by the analysis of sediment. The aspect "semi cloudy," recorded in most samples, appeared to be related to the aforementioned elements (semen, mucus, lymph, yeast, etc.) when the sediment was analyzed. The temporal record of urine appearance in this study may serve as a basis for future observations on *T. inunguis* urine.

3.2.3 Elements of the urinary sediment

Among the possible elements to be found in urine sediment, those detected included: epithelial cells, leukocytes, bacteria, crystals and hyphae. In none of the samples cylinders were observed.

3.2.3.1 Epithelial cells

Epithelial cells were recorded in all urine samples as rare, frequent (Figure 2) or numerous, regardless of gender. In males the most common finding was "rare" epithelial cells (n=84), whereas in females, frequent epithelial cells were recorded in most samples (n=50).

Epithelial cells are the squamous cell type most frequently observed in urine samples (Gürtler et al., 1987, Bossart et al., 2001). They may come from the bladder, urethra, renal pelvis or ureters, and may be present either singly or sometimes in small clumps (Bossart et al., 2001). In the Amazonian manatee they were present in both forms, mostly in urine samples from females. This may be due to physiological condition of increased shedding of the vaginal epithelium, especially if the animal is in heat (Gürtler et al. 1987; Kantek Garcia-Navarro).

Even in males this finding was observed in some samples mostly as "rare" cells. The detection of "numerous" cells was not considered worrisome, because it may be due to the normal shedding of urogenital tracts.

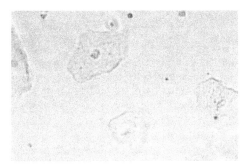

Fig. 2. Squamous epithelial cells found in urine samples from female *Trichechus inunguis* registered as "common".

3.2.3.2 White blood cells WBC's

White blood cells observed microscopically were presented in their degraded form (pus cells). The count of these elements in most examinations of males (n=93) was insignificant (1-3 per field) (Fig. 3), with three exceptions: a juvenile male in the month of June (Mapixari) and two subadult males in August (Anamã and Guarany), with 8, 10 and 8 pus cells by visual field, respectively. Most samples of females (n=58) also showed negligible amounts of pus cells (1-3 per field). However, adult females showed varying amounts of pus cells in their samples. High amounts of leukocytes were observed in the following tests: a female infant (Barreirinha) in August (30-35 pus cells by visual field), and an adult female (Cambá) in June (26 pus cells by visual field), September (18 pus cells by visual field) and October (30-40 pus cells by visual field).

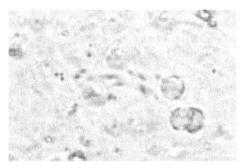

Fig. 3. Pus cells found in urine samples of male *Trichechus inunguis* (3 pus cells per field).

Leukocytes are found in normal urine, in relatively small amounts (Kantek Garcia-Navarro, 1996), as in the case of urine samples analyzed in this study. Leukocyte reaction (increased number of leukocytes in urine), may be due to inflammation somewhere in the urogenital apparatus. The large increase in the number of leukocytes (pus cells) in urine, a phenomenon observed in one infant and one adult female, is called pyuria (pus in the urine) (Kantek Garcia-Navarro, 1996) and in both of them, the increase of pus cells was accompanied by clouding of urine. However, these findings were not alarming, because the pyuria was not accompanied by cylinders, which could be indicative of more severe renal inflammation (Kantek Garcia-Navarro, 1996). In fact, cylinders were not found in any of Amazonian manatee's urine samples.

3.2.3.3 Bacteria

In male samples the occurrence of "bacteria" was recorded as absent, mild, moderate (Fig. 4) and increased. In most male urine samples (n=68) bacteria was absent, except in three cases: a young infant (Tuã), a subadult (Puru) and an adult (Yanomami) in December, when the "bacteria" was recorded as "increased" (bacteriuria). No females presented bacteriuria, and most of the samples (n=61) showed no bacteria.

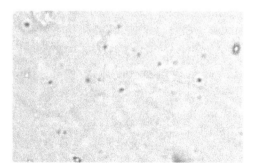

Fig. 4. Bacteria found in a sample of male *Trichechus inunguis* ("Bacteria" "moderate").

Bacteria in the urine may appear as small dots or dashes, both at rest, and in Brownian motion. Their presence in small amounts in urine is considered normal and can sometimes be due to environmental contamination (Bossart et al., 2001), or can derive from urinary meatus. Increased presence in urine may represent urinary tract infection, coming to form in more severe cases, elongated masses similar to cylinders. For the interpretation of cases in which the bacteria is increased in the urine, one should take into account the amount of leukocytes present. Only when both are found in increased amounts in urine, the existence of a severe infection may be considered (Kantek Garcia-Navarro, 1996).

In the rare cases where bacteriuria was observed in Amazonian manatee, this was not accompanied by increased leukocyte counts, which ruled out the possibility of severe infection (Kantek Garcia-Navarro, 1996). It is necessary to take into account that the presence of bacteria in the urine will be detected with nitrite reactive strips. In this study, the reactive strips showed low efficiency in the detection of nitrite, since in many samples, we obtained a positive value for nitrite, and yet, no bacteria were observed in them. Others, in turn, had bacteria when subjected to examination of sediment, but no nitrite was indicated by previous analysis by reactive strips. One possible explanation for this would be an accidental contamination of the sample prior to the study of urine sediment. This fact did not complicate diagnosis in our study, as we did not consider this result as an indicative of severe infection.

3.2.3.4 Crystals

The following crystals were observed microscopically in the urine samples of Amazonian manatees: amorphous phosphate crystals, triple phosphate crystals, amorphous urate crystals and calcium oxalate crystals (Figs. 5, 6, 7 and 8).

Fig. 5. Amorphous phosphate crystals found in urine samples of female *T. inunguis* recorded as "rare".

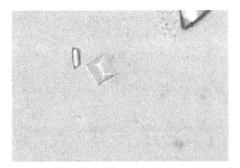

Fig. 6. Triple phosphate crystals found in urine samples of male *T. inunguis* recorded as "rare".

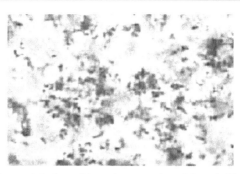

Fig. 7. Amorphous urate crystals found in urine samples of female *T. inunguis* recorded as "common".

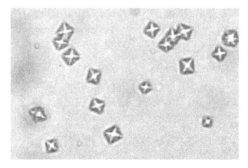

Fig. 8. Calcium oxalate crystals found in urine samples of *T. inunguis* recorded as "common".

In most examinations of males (n=78) amorphous phosphate crystals were not found, although the quantities observed in the samples varied widely and was also reported as rare, frequent and numerous.

In most samples of females (n=48) amorphous phosphate crystals were also not found. In the remaining ones, these crystals were recorded as rare or frequent. Unlike males "numerous" amorphous phosphate crystals were not found in any female samples.

The triple phosphate crystals were not recorded in most samples of males (n=82). When found they appeared as "rare", "frequent" or "numerous". Most samples of females (n=73) presented no triple phosphate crystals. However, in some samples they appeared as "rare," "frequent" or "numerous."

Except for one infant male calf (Piraí), which in the months of June and July had "numerous" and "frequent" amorphous urate crystals, these crystals were not registered in any other sample from males. There was also no evidence of amorphous urate crystals in the samples of most females (n=75).

Finally, calcium oxalate crystals were observed in only four samples of males: as "rare" in a young infant (Piraí) in June and in a juvenile (Tapajós) in September, and as "frequent" in a juvenile male (Tapajós) and in a subadult (Anamã), both in August. The remaining samples of males, as well as all the females' samples, showed no calcium oxalate crystals.

Crystals are formed by minerals present in urine, so it may appear amorphous or crystalline. The crystals can be either in the bladder of the animal (*in vivo*), as *in vitro*. In fact, the presence of crystals is usually nonspecific, with no diagnostic value, and it has to take into account other factors such as diet, patient medication, urine specific gravity, time and storage conditions of the sample, and especially urinary pH to obtain a better interpretation of its occurrence in the urine (Kantek Garcia-Navarro, 1996).

The pH is actually a regulator of crystal formation, as shown in Table 2, which discriminates the crystals observed in Amazonian manatee urine and the respectively urinary pH related to their occurrence. In all cases, the type of crystal was recorded in urine with a pH suitable to its appearance.

Crystal	Urinary pH
Phosphate amorphous	Alkaline
Triple phosphate	Alkaline, neutral or slightly acid
Amorphous urate	Acid
Calcium oxalate	Acid, neutral or alkaline

Table 2. Crystals observed in Amazonian manatee urine and urinary directly related to their occurrence (Adapted from Benjamin (1976)).

Amorphous phosphate crystals actually consist of clusters of fine granules which, although being very small, could be distinguished from bacteria by their refringence. An increase in number in urine may be due to retention of urine in the bladder, enlarged prostate and chronic pyelitis. Triple phosphate crystals (or struvite crystals) emerge from the alkaline fermentation of urine, which can occur both before and after urination. When found in urine freshly collected it may be indicative of urine retention in the bladder (as in chronic cystitis), paraplegia, enlarge of prostate size or chronic pyelitis.

These crystals can also be observed in feline urologic syndrome, or may indicate the presence of a struvite urolithiasis (Kantek Garcia-Navarro, 1996). They are also commonly found in the urine of horses (M. S. Matos & P. F. Matos, 1995). There are records of incidental findings of these crystals in ringed seal (*Pusa hispida*), elephant seal (*Mirounga angustirostris*), Weddell seal (*Leptonychotes weddellii*), *T. truncatus* and humpback whale (*Megaptera novaeangliae*) by some authors (Gulland et al., 2001). The triple phosphate crystals can present the form of short or elongated prisms, resembling at times the cover of a coffin box or the roof of a house. They were observed in the form of a leaf or fern (penalty) (Kantek Garcia-Navarro, 1996). In urine samples of the Amazonian manatee, they presented themselves in the form of box coffin cover. The observation of phosphate crystals in the amorphous and triple phosphate tests did not follow any pattern, but their record may be used as a basis for future studies.

Amorphous urate crystals appear as clusters of small yellow-brown granules, which may also suffer crystallization and appear as colourless fine needles, arranged in the form of pulleys or sheaves (M. S. Matos & P. F. Matos, 1995; Kantek Garcia-Navarro, 1996). By observation of amorphous urate crystals in few samples of both males and females, it was noted that they are hardly found in *T. inunguis* urine, which is mainly due to the fact that the urine of these animals is normally alkaline, and these crystals are evidenced in acidic urine. In this study, the few tests at those amorphous urate crystals were detected had pH 6.0.

Easily identified by taking the classic "envelope letter" form, the crystals of calcium oxalate are considered a normal finding in urine, especially after eating tomatoes, garlic, oranges and asparagus. They are also commonly found in the urine of horses (M. S. Matos & P. F. Matos, 1995). However when observed in increased numbers in the urine they may be related to diabetes mellitus, liver disease, heart, or lungs (Kantek Garcia-Navarro, 1996). The occurrence of this type of crystal was recorded by De Monte & Pilleri (1972) in the urine of *I. geoffrensis* and *P. gangetica*. As just four samples of Amazonian manatees showed to have these crystals, it seems that they are not normally found in the urine of this species. These rare cases were probably due to factors related to ingestion of tomatoes, which was part of the diet of these animals in the days before the exams, since the animals did not show any apparent symptoms of above conditions.

3.2.3.5 Blood cells (RBCs)

In most microscopic examinations of both males (n=105) as females (n=68), red blood cells were not seen. However, these were present in some samples (Fig. 9). Unusually, the tests of an adult female (Cambá) showed 80-100 red cells per field in July and more than 10 red cells per field in the months of August and December.

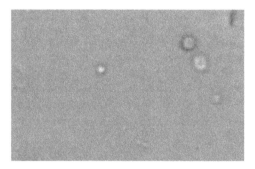

Fig. 9. Red blood cells found in urine samples of male *Trichechus inunguis*.

As previously mentioned, the results obtained qualitatively by reactive strips were discarded and the study of red blood cells by microscopic analysis of the sediment was considered the most efficient method. In females, the appearance of red blood cells in urine, if any, was in very small quantities, and even in samples where the number of red blood cells was found to be slightly higher (15 to 20 erythrocytes per field), these findings do not represent a worrying result, since their occurrence was not accompanied by symptoms related to disease previously described as causes of haematuria. Moreover, the presence of red blood cells in these samples failed to give a reddish-brown color to the samples as expected for samples containing higher amounts of this cell type (Kantek Garcia-Navarro, 1996). In males, when there was detection of red blood cells, the cellular amounts of this element were too small serve as diagnosis of a disorder in these animals.

3.2.3.6. Hyphae

In most tests, both in males (n=87) and females (n=58), hyphae were not observed microscopically. In samples that contained the hyphae they appeared in small quantities (Fig. 10).

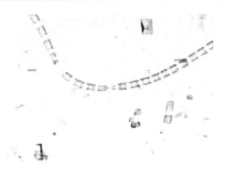

Fig. 10. Hyphae found in urine samples of male *Trichechus inunguis*.

In general, hyphae are the result of material contamination. However, true infections have been recorded in domestic animals (Kantek Garcia-Navarro, 1996). None of the Amazonian manatee's urine samples showed high enough amounts of hyphae to suggest a serious infection by fungi. Even samples that contained hyphae showed very small amounts, which may have been due to contamination of the sample during urine collection and/or analysis.

4. Conclusions

The qualitative analysis by reactive strips proved to be useful for urinalysis in the Amazonian manatee in captivity for the parameters pH, density, bilirubin, urobilinogen, ketones, protein and glucose. On the other hand, the analysis of: blood (erythrocytes), nitrite and leukocytes needs to be confirmed by microscopic analysis of the sediment.

Given the constancy of the results of glucose, bilirubin, urobilinogen, ketones and reactive protein measured by the strips, they appear to be the minimum expected normal values of these substances in the urine of *T. inunguis*. Despite the qualitative nature of results obtained by reactive strips, from a practical standpoint, they can be applied routinely to monitor these parameters.

The urinary sediment showed no cylinders, suggesting that the animals analyzed had no illnesses related to their presence. The observed elements that could indicate some pathology, such as leukocytes or bacteria, were present just in small quantities, or accompanied by casts (in the case of leukocytes) or pyuria (in the case of bacteria), ruling out the possibility of disease in sampled animals. The temporal record of the results of other elements of urinary sediment, as well as physical aspects of color and appearance will be the basis for future comparative analysis on urinalysis animal.

Results reported in this study may help monitor the health situation of *T. inunguis* both in captivity and in the wild, as well as being referential data for urinary parameters in related species. This investigation, by furnishing physiological knowledge to assess the health condition of these animals, provides basis for management and conservation of this vulnerable species.

Additionally, knowledge of urinary compounds levels, as well as their patterns, provided by this study underlies actions for Amazonian manatee conservation, such as the reintroduction of rehabilitated animals after stranding or bycatch, since it enhances the possibility of success of such procedures.

Finally, we emphasize the need for future research on variations in some urinary components levels in response to experimentally induced physiological stress. These results can be compared with those obtained in this study in order to try to understand the mechanisms of homeostatic self-regulation of this currently threatened species.

5. Acknowledgments

We would like to thank Sarita Kendall, Hermes Schmitz and Alexander C. Lees that reviewed the English. Special thanks for Instituto Hospital Tropical, Manaus, Amazonas, Brazil by providing us the logistical support to carry out urine analyses.

6. References

Ayres, J.M. & Best, R. C. (1979). Estratégias para a conservação da fauna amazônica. Acta Amazônica, v. 9, n.4, p. 81-101.

Best, R. C. (1982). A salvação de uma espécie: novas perspectivas para o peixe-boi da Amazônia. Revista IBM, n.14, p. 1-9.

Best, R. C. (1984). *Trichechus inunguis,* vulgo peixe-boi. *Ciência Hoje,* 2(10):66-73

Bossart, G. D., Reidarson, T. H., Dierauf, L. A. & Duffield, D. A. (2001). Clinical Pathology. *In:* Dierauf, L. A.; Gulland, F. M. D. (Eds.) *CRC Handbook of Marine Mammal Medicine. 2nd. Ed.* CRC Press, Boca Ratón, EUA. p.383-436

Costa, L. P.; Leite, Y. L. R.; Mendes, S. L.; Ditchfield, A. D. (2005). Conservação de mamíferos no Brasil. Megadiversidade, n. 1, p. 103-112, julho.

De Monte, T. & Pilleri, G. (1972). Hematological, plasmatic and urinary values of three species of cetaceans belonging to the family Platanistidae [*Inia geoffrensis* (de Blainville, 1817); *Platanista indi* (Blyth, 1859); and *Platanista gangetica* (Roxburgh, 1801)]. *Revue Suisse de Zoologie,* 79,1(7):235-252

Gulland, F. M. D., Lowenstine, L. J. & Spraker, T. R. (2001) Noninfectious Diseases. In: CRC Handbook of Marine Mammal Medicine. Dierauf, L. A.; Gulland, F. M. D. (eds.) 2nd Ed CRC Press. p. 521-537

Gürtler, H.; Ketz, H.-A; Kolb, E.; Schröder, L. & Seidel, H. (1987). *Fisiologia Veterinária. 4ª Ed.* Editora Koogan, Rio de Janeiro. 612p.

Hilton-Taylor, C. (Comp.) (2000). *2000 IUCN Red List of Threatened Species.* IUCN/SSC, Gland, Suíça/Cambridge, Reino Unido. http://www.iucn.org/redlist/2000

Junk, W. J.; Da Silva, V. M. F. (1997). Mammals, reptiles and amphibians. Ecological Studies, Berlim, v. 126, p. 409-417.

Kantek Garcia-Navarro, C. E. (1996). *Manual de Urinálise Veterinária.* Livraria Varela, São Paulo. 95p.

Maluf, N. S. R. (1989). Renal anatomy of the manatee, *Trichechus manatus*, Linnaeus. *American Journal of Anatomy,* 184:269-286

Manire, C. A., Walsh, C. J., Rhinehart, H. L., Colbert, D. E., Noyes, D. R. & Luer, C. A. (2003). Alterations in blood and urine parameters in two Florida manatees (*Trichechus manatus latirostris*) from simulated conditions of release following rehabilitation. *Zoo Biology,* 22:103-120

Marvulo, M. F. V., Da Silva, V. M. F., Martin, A. R., D'Affonseca Neto, J. A., Rosas, F. C. W., Nascimento, C. C., Morais, Z. M., Vasconcelos, S. A., Ferreira, J. S. N. & da Silva, J. C. R. (2003). Serosurvey for antibodies against *Leptospira sp.* and *Brucella sp.* in free living amazon river dolphins (*Inia geoffrensis*) and captive amazonian manatees (*Trichechus inunguis*). *Annals of 15th Biennial Conference on the Biology of Marine Mammals.* Greensboro, North Carolina, EUA. p.104

Matos, M. S. & Matos, P. F. (1995). *Laboratório Clínico Médico-Veterinário. 2ª Ed.* Editora Atheneu, Rio de Janeiro. 238p.

MMA. (2001). Mamíferos Aquáticos do Brasil: Plano de ação, Versão II.-2.ed.rev., aum.- Brasília: Instituto Brasileiro do Meio Ambiente e dos Recursos Naturais Renováveis – IBAMA, 102p.

Ortiz, R. M. (2001) Osmoregulation in marine mammals. The Journal of Experimental Biology. Printed in Great Britain. 204, 1831–1844.

Pantoja, T.M.A., Rosas, F.C.W., Da Silva, V.M.F. & Dos Santos, A.M.F. (2010). Urinary parameters of *Trichechus inunguis* (Mammalia, Sirenia): reference values for the Amazonian Manatee. *Braz. J. Biol.* [online]. 2010, vol.70, n.3, pp. 607-615.

Rodriguez, Z.M., Da Silva, V.M.F. & D'Affonseca Neto, J.A. (1999). Teste de fórmula láctea na alimentação de filhotes órfãos de peixe-boi da Amazônia (*Trichechus inunguis*). In : Fang, T.G.; Montenegro, O.L.; Bodmer, R.E. (Eds.). Manejo y Conservación de Fauna Silvestre en América Latina. Bolivia: Instituto de Ecología. p. 405-408.

Rosas, F. C. W. (1991). Peixe-Boi da Amazônia, Trichechus inunguis (Natterer, 1883). In: Cappozzo, H. L.; Junin, M. (Eds.). Estado de conservación de los mamíferos marinos del Atlântico sudoccidental informes y estudios del Programa de Mares Regionales del Programa de las Naciones Unidas para el Medio Ambiente (UNEP), ONU, ROMA, n. 138, p. 178-181.

Rosas, F. C. W. (1994). Biology, conservation and status of the Amazonian manatee *Trichechus inunguis. Mammal Review,* 24(2):49-59

Rosas, F. C. W. & Pimentel, T. L. (2001). Order Sirenia (manatees, dugongs, sea cows). *In*: Fowler, M. E.; Cubas, Z. S. (Eds.). *Biology, Medicine, and Surgery of South American Wild Animals.* Iowa State University Press, Ames, EUA. (31):352-362

Schweigert, F. J. (1993). Effects of fasting and lactation on blood chemistry and urine composition in the grey seal *(Halichoerus grypus). Comparative Biochemistry and Physiology,* 105A (2):353-357.

Strasinger, S. K. (1996). *Urinalysis and Body Fluids. 3rd Ed.* F. A. Davis Co., Filadélfia, EUA. 233p.

Trujillo, F.; Kendall, S.; Orozco, D.; Castelblanco, N. (2006). Manatí amazônico *Trichechus inunguis,* p. 167. In: Rodríguez - M., J.V.; Alberico, M.; Trujillo, F.; Jorgenson, J. (Eds.). Livro rojo de los mamíferos de Colombia. Série livros rojos de especies amenazadas de Colombia. Conservación Internacional Colombia, Ministerio de Ambiente, Vivienda y Desarollo Territorial. Bogotá, Colombia, 2006.

Trujillo, F.; Alonso, J. C.; Diazgranados, M. C.; Gómez, C. (2008). Fauna acuática amenazada en la Amazonía colombiana: análisis y propuestas para su conservación. Bogotá: Fundación Omacha, Fundación Natura, Instituto Sinchi, Corpoamazonía, 125p.

Vianna, J-A.; Dos santos, F.R.; Marmontel, M.; De Lima, R.P.; Luna, F-O.; Lazzarini, S.M.; De Souza, M.J. (2006). Peixes-boi: esforços de conservação no Brasil. Ciência Hoje, v.39, n.230, p. 32-37.

5

Global Efforts to Bridge Religion and Conservation: Are They Really Working?

Stephen M. Awoyemi[1], Amy Gambrill[2],
Alison Ormsby[3] and Dhaval Vyas[4]
[1]*Tropical Conservancy,*
[2]*International Resources Group,*
[3]*Eckerd College,*
[4]*Alverado Way, Decatur, Georgia,*
[1]*Nigeria,*
[2,3,4]*USA*

1. Introduction

1.1 The biodiversity crisis and the need for global collaboration

It has been close to half a century since Rachel Carson's *Silent Spring* was published — a book that marked the beginning of the modern environmental movement — yet the biodiversity crisis has increased to staggering proportions. The International Union for the Conservation of Nature (IUCN) estimates that the current species extinction rate is between 1,000 and 10,000 times higher than natural rates (IUCN, 2011). According to the United Nations Food and Agriculture Organization's (FAO) Global Forest Resources Assessment 2005, forests have disappeared in 25 countries and deforestation clears about 12 million hectares annually, including six million hectares of primary forests in Latin America, South-East Asia and Africa (FAO, 2006). Twenty percent of coral reefs have been destroyed and 30 percent damaged due to destructive fishing practices, pollution, disease, coral bleaching, invasive alien species and tourism (Millennium Ecosystem Assessment, 2005; Wilkinson, 2008). Driven by habitat destruction, invasive species, pollution, human overpopulation and over harvesting (all human-centred activities), the global loss of biodiversity is advancing at an unprecedented rate with an extinction of 150 species occurring daily (Chen, 2003; Sigmar, 2007). The United Nations Environment Programme (UNEP) reports that the natural systems that support the world's economy are at risk of collapse due to increasing biodiversity loss: declining fish stocks, deforestation and soil erosion (Murray, 2010). However, response to the crisis has been slow. For instance, the Convention on Biological Diversity's target to reduce the rate of biodiversity loss by 2010 was missed (Butchart et al., 2010).

The paucity of response by governments, corporate organizations and the public may be attributed to several factors, which include but are not limited to modicum of biodiversity awareness by humans and competing factors such as poverty alleviation, unending growth in consumption and human numbers, war, terrorism, healthcare, political and economic challenges. Novacek (2008) posited that the public is a primary force for generating momentum to drive governments and business corporations in dealing with the biodiversity crisis. Although the public is made up of a wide range of groups in different social positions – women, men, adults, children, old, workers, professionals, the wealthy, etc. – all with differing and sometimes competing interests, only some of these groups can reach decision-makers. To have an effect, groups must be organized and be able to reward and admonish. The biodiversity crisis is global and connected to all humanity, although in different ways: some benefit from destroying biodiversity; others are hurt by such destruction, materially or psychologically; and for most, the immediate effects of loss are distant. Since the crisis knows no jurisdiction, geopolitical zones or national boundaries, harnessing public interest and involvement throughout the world could present an immense opportunity for biodiversity conservation. Human history documents the centrality of religion in regulating peoples' actions, including ecologically important behaviour, and the role of religious-based mobilization in creating policy changes (Rappaport, 1974, 1976, 1999; Wilson, 2002). In this light, it is important to better understand the potential of religion-based public support for conservation of biodiversity and the environment. Approximately four billion people in 125 countries listed in Conservation International's (CI) biodiversity hotspots are affiliated with one of 11 mainstream faiths (Bhagwat & Palmer, 2009), and many at the same time practice indigenous traditional religions. There is a need to assess past and current efforts to link religion and conservation to affirm their practicality and, if appropriate, suggest an effective path forward. This review will therefore: first, examine the rationale for the bridging of religion and conservation; second, provide a brief history of the efforts to bridge religion and conservation; third, highlight projects in different parts of the world that bridge religion and conservation showing their outcomes; and fourth, analyze successes and challenges in religion-based mobilization on behalf of conservation.

2. Why religion-based mobilization could strengthen conservation

Although significant, efforts by scientists and conservation organizations to conserve biodiversity have proven insufficient in curbing biodiversity loss. What, therefore, can the conservation community do to engage groups to foster behavioural (both individual and structural/institutional) changes for biodiversity conservation? This change would need to pervade every community worldwide and cascade across generations in perpetuity to secure the future of biodiversity. All individuals have values, attitudes, motivations and judgments, and these are often based in and sanctified by religious beliefs. Religion is a powerful influence on human behaviour, guiding thought processes and daily living for over 80 percent of the global population (Rappaport, 1979, 1999; Higgins, 2011). Targeting a person's deeply entrenched paradigm such as a religious worldview may be more effective in persuading people to make changes in daily behaviour, including engaging in activity to influence institutions on behalf of biodiversity. According to Gambrill (2011), Muslim fishers in Misali Island, west of Pemba, Tanzania, once threatened important turtle nesting sites and delicate coral slopes through dynamite fishing. Efforts by government and environmental agencies to educate the populace proved ineffective until the Islamic Foundation for Ecology and Environmental Sciences (IFEES) conducted two environmental ethics workshops based on the Qu'ran, in 1998 and 2001. The central message from the Imams to the madrassa teachers and fishermen leaders was that dynamite fishing was illegal according to Islam, and the fishermen responded by ending this practice immediately.

Faith communities comprise the largest social organizations in the world, spanning national and other divisions. They possess centuries of experience in offering practical and authoritative guidance in daily living that could benefit biodiversity. A World Bank report (2006, page 1-2) identifies three paths for religious influence: "1. They can teach about the environment and natural systems upon which life depends; 2. They can provide leadership in initiating practical environmental projects and 3. They can seek to persuade their members that each individual has a moral obligation to contribute in some way to conservation, and can provide guidance on how to pursue environmental management objectives". They can also instigate political action on behalf of biodiversity. In the United States for example, a network of churches was able to mobilize politically and help stop the U.S. Congress from weakening the Endangered Species Act (Barcott, 2001). Also, the Religious Bodies Environmental Network (RELBONET) in Ghana has lobbied the Government of Ghana to work on climate change issues (Otabil, 2010). Rolston (2010) opines that religious faith can make a unique contribution to environmental policy. He argues that scientific reasoning is able to give only partial and value-free guidance but religious faith and communities can, and have already begun to offer what science lacks: "a value-laden, unified understanding of creation, humankind and our obligations as stewards of the earth".

The potential of religion to support conservation goals has elicited significant interest and generated dialogue since the World Wildlife Fund's (WWF) summit on religions and conservation in Assisi, Italy in 1986 (there are also historically deeper examples such as Buddhism teaching compassion for all things and St. Francis of Assisi opposing animal cruelty). However, the history of religions reveals many aspects that are antithetical to conservation and also a great heterodoxy, with different traditions within and among religions concerning the natural world. The dominant tendency in many faiths is anthropocentric such as the Abrahamic (Judaic, Christian and Islamic) belief that humans are stewards of Yahweh-God-Allah's creation which has been given to us as a gift. Other traditions such as Jainism are biocentric and hold that every being – animal, plant or human – has a soul and should be treated with respect (Hall et al., 2009). These are complex waters for conservationists to navigate.

In seeking cooperation with religious institutions, it helps to distinguish between religious scripture and practice that directly protects some natural places as sacred and scripture and practice guides or prescribed behaviours that affect nature. We will look at each in turn.

First, common ground between religions and biodiversity protection is found throughout the world in the form of sacred natural sites and religious-based behavioural control systems (Dudley et al., 2005). Religions such as Animism, Hinduism, Buddhism and African Traditional Religion[1], among others, protect sacred species, groves, and forests. For example, one African Traditional Religion in Osun State, Nigeria protects the Osun-Osogbo sacred groves; they were named by the United Nations Educational, Scientific and Cultural Organization (UNESCO) as a World Heritage Site in 2005 (UNESCO, 2011). Unique flora such as *Adansonia digitata, Bombax buonopozense, Newbouldia laevis* and *Melicia excelsa* found in the Osun-Osogbo sacred groves are preserved by the locals based on beliefs of the sacredness of these species (Babalola, 2009). The Baka (pygmy) people, who inhabit the

---

[1]African Traditional Religion means indigenous religious beliefs and practices of Africans. Source: Awolalu (1976)

forests of Lobeké National Park in southeast Cameroon, adhere to a complex faith system that includes the adoption of a personal deity in adolescence and the worship of particular groves and trees inside the forest believed to be of high spiritual value. These sacred places are carefully protected (Dudley et al., 2005). Societies conserved sacred sites long before the emergence of modern protected areas. This is probably the oldest method of habitat protection on earth and still forms a large and unrecognised network of sanctuaries around the world (Dudley et al., 2005; Verschuuren et al., 2010). The WWF and Alliance of Religions and Conservation (ARC) report (Dudley et al., 2005) included a partial survey of 100 protected areas around the world that included sacred sites as well as sacred areas outside of protected areas that have high conservation values. The study revealed that links between faiths and protected areas are neither unusual nor limited by either geography or faith; rather the links are substantial and pervasive.

Second, many religious believers look to local and global religious authorities for guidance not only concerning larger purposes and meaning but for how to live their daily lives in accordance with their larger purposes. For this reason, religious leaders have the capacity to convey to their believers how their values can direct their behaviour toward the natural world in ways that conserve biodiversity. Religious leaders whose influence extends globally can transcend national boundaries, which are often a stumbling block in conservation. Biodiversity hotspots identified by CI (Bhagwat et al. 2011a) are located in countries in which 70 percent of the population on average adheres to a religion. The power of religion to check destructive behaviour in the face of challenges ranging from political instability and conflict to poverty and lack of empathy for other creatures is significant, and if it can be mobilized in support of biodiversity it would be a noteworthy achievement (Rappaport, 1999).

Because many religions are concerned with the human environment, not biodiversity, there is not yet consensus on what to do about biodiversity conservation and whether it should be a priority. When conservationists seek support, they need to know their audience.

3. A brief history of efforts to bridge religion and conservation

The earliest major effort to bridge religion and conservation was in 1986 during the celebration of the 25th anniversary of WWF at the Basilica of St. Francis in Assisi, Italy. The meeting was convened by His Royal Highness Prince Philip, Duke of Edinburgh, then President of WWF/International. It brought together 800 people and generated the *Assisi Declarations,* which included extractions of environmental ethics from five mainstream faiths: Buddhism, Christianity, Hinduism, Islam and Judaism (Sponsel, 2007). The meeting also initiated the Network on Conservation and Religion sponsored primarily by the WWF. In 1995, Prince Philip hosted another summit, held in Windsor Castle, England. Nine major world religions – the Bahá'í Faith, Buddhism, Christianity, Hinduism, Jainism, Judaism, Islam, Sikhism and Taoism – along with key officials from several major secular institutions gathered to discuss how the world's religious communities might become engaged in environmental conservation (One Country, 1995). Key results from the summit included a dramatic commitment by each of the faith communities to heighten their efforts at furthering the cause of conservation within their own fold, along with a new level of interfaith cooperation and agreement (One Country, 1995). The summit also marked the evolution of the Network on Conservation and Religion into a more self-governing group called the Alliance of Religion and Conservation (ARC) (One Country, 1995). ARC is a secular body that assists the world's major religions in developing their own environmental programmes. The religious institutions work according to their core teachings, beliefs and practices. ARC currently works with 11 major faiths worldwide (ARC, 2011a).

Another cooperative effort is Harvard's Centre for the Study of World Religions. John Grim and Mary Evelyn Tucker (now at Yale University) started a series of conferences on religion and ecology from 1996-1998 bringing together over 800 environmentalists and international scholars of the world's religions (FORE, 2011). The conference "Science, Religion and the Natural World" was held May 11-14, 2000 at Yale University (Yale University, 2011) and resulted in the book "The Good in Nature and Humanity: Connecting Science, Religion, and Spirituality with the Natural World" (Kellert and Farnham, 2002). In 2006, the Forum on Religion and Ecology (FORE) at Yale University was established; it is the largest international multi-religious project of its kind (FORE, 2011). There are also several international faith organizations specifically dedicated to the environment, including the Association of Buddhists for the Environment, A Rocha, Islamic Foundation for Ecology and Environmental Sciences, Coalition on the Environment and Jewish Life, Evangelical Environmental Network and The Blessed Kateri Tekakwitha Conservation Centre (formerly the Catholic Conservation Centre). Furthermore, there is cooperation among different faith groups based on their common interest in conserving life on earth. Cooperation may be informal or formal, including organizations such as Interfaith Power and Light, World Council of Churches, All Africa Council of Churches, and Southern African Faith Communities Environment Institute. Secular organizations such as WWF, the Wilderness Society, United Nations Development Programme (UNDP), World Bank, CI, UNEP and IUCN all currently have or have had outreach programmes to bridge religion and conservation.

In November 2009, ARC and UNDP joined with 31 faith traditions to launch and celebrate their long-term commitments for a living planet. The collaboration resulted in the Guide to Creating Seven Year Plans (2010-2017) which centres on environmental action, reflection and thought. The Guide employed seven pivotal areas in which faiths possess influence – ranging from investments, through partnerships and media to education and celebration – offering ideas on how each can utilize their strengths to take specific steps toward increased biodiversity protection (ARC, 2011b).

4. A global review of specific projects that bridge religion and conservation

4.1 Conservation International (CI) Indonesia

CI Indonesia implemented the Islamic Boarding School and Conservation Project from December 2004 to June 2005 with an aim to facilitate, deepen and raise the awareness about religious arguments for forest and biodiversity protection and stewardship in Indonesia, and the linkage between conservation, human welfare and poverty alleviation (CI, 2005). The target audience of the project consisted of students, teachers, surrounding community members, religious leaders, pesantren (Islamic boarding school) scholars, individuals and institutions.

Although a relatively short project, the main activities were:

- Production of a book about nature conservation: 2,000 copies were distributed for free to schools, religious organizations and other stakeholders.
- Creation of a small grants program for Islamic boarding schools for activities relating to biodiversity conservation.
- Workshops that brought together participants to promote nature conservation in daily life and to link the Islamic community with environmental conservation efforts.

Lessons learnt included:

- An enthusiastic response concerning new publications related to conservation and Islam, namely *"Konservasi Alam Dalam Islam"* (Nature Conservation in Islam) and an Indonesia Forest Media and Campaign (INFORM) report: Fiqh Al-biah. The new approach to providing in-depth information on traditional Islamic wisdom and teachings concerning nature conservation changed peoples' perspective and their way of living with nature. Students from the pesantren who were given copies of *Konservasi Alam Dalam Islam* for example, coupled with the administration of a small grants programme for reforestation activities, planted 2825 trees from 15 species and conducted field activities related to forest and biodiversity conservation.
- Use of positive religious teachings (particularly Islamic teachings) as a soft approach to promoting conservation can lead to easier acceptance of environmental messages by communities around protected areas;
- Mosques and Islamic boarding schools can provide a practical environment for promoting conservation and environmental awareness; and
- There is little resistance on the part of the Islamic communities in Indonesia to bridging conservation and religious teachings, and this is seen as a positive correlation between religion, humanity and nature.

4.2 World Bank[2]

4.2.1 Papua New Guinea

In Papua New Guinea, the World Bank Faith and Environment programme supported construction of a centre for theological involvement in forest conservation, adjacent to a nature reserve in the Eastern Highlands province. The centre is also involved in the development of theological literature covering conservation issues. It started with a meeting of Christian leaders in Goroka in 2003, which led local churches to proclaim their commitment to care for the environment in Papua New Guinea. In the meeting, the idea of biblically based environmental stewardship was introduced for the first time for many attendees. Subsequent to the meeting, the World Bank together with TearFund Australia and World Council for Missions funded the production of a handbook on theology and the environment, entitled *Christians Caring for the Environment*, compiled by Evangelical Alliance; the book was endorsed by the Catholic archbishop of Papua New Guinea and now serves as a model for similar publications being prepared in Africa and the Pacific.

---

[2]Highlights of the following conservation projects with religious connections are from the World Bank (2006) report entitled "Faith and the Environment - World Bank Support 2000-2005".

4.2.2 South Africa

Working for Water, KwaZulu-Natal Department of Agriculture and Environmental Affairs, ARC, and South African faith-based organizations collaborated in establishing an innovative pilot project manufacturing affordable coffins made from invasive tree species. This partnership, with technical assistance from the South African Nursery Association, will also fund the growing of indigenous plants used for the restoration of areas cleared of invasive species; the planting of native trees also honours the memory of the dead.

4.2.3 Cambodia

In Cambodia, the World Bank in partnership with ARC supported Mlup Baitong, an NGO that provided environmental education and training. Target audiences during the five-year programme included 14 pagodas in rural areas in Kampong Speu and Kampong Thom provinces. Monks living in these pagodas as well as achars, nuns and villagers in the surrounding districts received instruction. In addition, these pagodas became promoters of sustainable development models for other neighbouring pagodas and all villages in the area. The Ministry of Environment, Department of Education, Provincial Environmental Department, National Park authorities and local officials also cooperated in the project. Monks and achars gave training on Buddhism and the environment and the practical application of these skills to villagers and visitors through lectures, workshops and closed-circuit radio programmes on Buddhist holy days. Some 55 workshops for monks were held including 450 village lectures given by the monks and two provincial network meetings for all participating monks. Tree nurseries are now well-established in the pagodas, and some of these seedlings are planted on pagoda grounds, while others are donated to the community. As part of the school environment programme, tree nurseries and compost bins have also been established in eight schools and over 1,000 trees have been planted with monks organizing seedling ordination ceremonies and tree planting days.

4.2.4 All Africa Council of Churches and Alliance of Religions and Conservation Collaboration

ARC collaborated with the All Africa Council of Churches to research the level of current activity connected to the environment and areas where churches in Africa would like to develop programmes. The study revealed that 76 respondent churches and councils viewed ecological sustainability as a commission by God to the church. In this view economic development must be managed in a manner that protects the environment; this is seen as "co-working with God" and caring for God's creation. A high proportion of the respondent churches had projects connected to environmental management, with forestry and reforestation pinpointed as key areas.

4.3 A Rocha[3]

4.3.1 Ghana

A Rocha Ghana has been working since 2005 with the Collaborative Resource Management Unit of Mole National Park, district assemblies and surrounding communities in building

---

[3]For more information on these examples of A Rocha conservation projects, see Sluka et al. (2011).

consensus and support for sustaining the ecology of Mole National Park. A Rocha Ghana's engagement with surrounding communities mobilized support in part by appealing to their belief systems, linking conservation concerns to existing beliefs. Faith-based environmental messages were delivered to both Christian and Islamic groups in twelve communities by A Rocha Ghana. Also a Community Resource Management Area (CREMA) with a constitutional and legal framework has been established, giving these communities the authority and incentives to sustainably manage and conserve the local natural resources. The results of this project include a deeper commitment from local communities to use natural resources sustainably, demarcation of 681 km² of communal land designated as a core management area with access and use regulations supported with by-laws, and a reduction in illegal hunting and the return of wildlife to areas near CREMA villages. Many individuals benefited directly from natural resource-based enterprises such as beekeeping and local communities generated US$7,100 in 2009. The Murugu-Mognori CREMA, which is 268 km² receives an annual revenue of US$4,000 from the "Mognori Eco-Village" ecotourism initiative. Additionally, 72 households have enjoyed a 220 percent increase in household income from beekeeping alone and 30 households now enjoy additional income and improved nutrition from vegetable production.

4.3.2 Kenya

A Rocha Kenya was established in 1999 and works to protect Important Bird Areas on the north Kenya coast, especially the Arabuko-Sokoke Forest and Mida Creek. A Rocha Kenya formed ASSETS (the Arabuko-Sokoke Schools and Eco-Tourism Scheme) with a goal to conserve the forest and concomitantly allow families to benefit directly from its conservation by raising funds for community members' secondary school fees through eco-tourism, thereby reducing one of the drivers of illegal logging. Project results as of September 2010 show that 378 children attended secondary school on ASSETS bursaries and 144 have graduated. To make certain that parents and students know that forest conservation is funding their schooling, both are engaged in environmental education activities, water conservation initiatives and litter clean-ups; they are also asked to pledge to abstain from illegal logging and poaching. ASSETS beneficiaries are also provided with free seedlings to grow their own woodlots to reduce pressure on the forest and tree nurseries have been established in local schools. A monitoring plan was developed in 2007 which to date has shown that parents of ASSESTS students show a strong protective attitude toward the forest, a greater general environmental awareness and better understanding of the connection between forest conservation and their children's access to education.

4.4 The International Small group and Tree planting program (TIST)

4.4.1 Kenya, Tanzania, Uganda, Honduras, Nicaragua and India

In 1998, at the invitation of the bishop and his wife, a team of U.S.-based missionaries came to the Anglican Diocese of Mpwapwa in Tanzania to hold a seminar with subsistence farmers to discern the best practices for small groups based on servant leadership. The following year, another small group training seminar resulted in the farmers' development of a vision to reforest their land, eradicate famine, initiate health interventions and start new, small groups over the following year.

In response to the Tanzanian farmers' vision and to create a sustainable income for the farmers with a self-sustaining program, The International Small Group and Tree Planting Program (TIST), led by Clean Air Action Corporation, was born at the end of 1999. It was based on the God-centred small group best practices – including rotatation of leadership, servant leadership, use of co-leaders, agreement to a covenant and accountability to one another – developed at the seminars, and addressed the overwhelming deforestation, drought and famine in the area through the church. By 2000, TIST was open to people of all religions. TIST's Board wanted whole communities to benefit (www.tist.org).

TIST implementation in Kenya, Tanzania, Uganda, Honduras, Nicaragua and India has resulted in more than 9,000 small groups with over 60,000 members planting 11 million trees. In 2011, TIST was validated and verified for their work in Kenya through the Verified Carbon Standard (VCS) and also by the Climate, Community & Biodiversity Standards (CCBS). TIST was the first carbon offset program in the world to have achieved dual certification. This strength and growth is founded on principles developed by faith communities sharing and acting on their vision of stewardship and action for the planet and servanthood to each other (*pers. comm.* Vannesa Hennecke, August 2011).

A key lesson learned is that faith communities and partners provide strong entry points (*pers. comm.* Vannesa Hennecke, July 2011). TIST's program was implemented in Kenya through Kenya's Forest Service, but most participants today are members of Catholic, Pentecostal and Anglican Churches. In Uganda, the program began with the Anglican Church. In Nicaragua and Honduras, TIST partnered with Catholic Relief Services to introduce the program.

4.5 Flora and Fauna International (FFI) and Uganda Wildlife Authority

4.5.1 Uganda

Rwenzori Mountains National Park is home to the Bakonjo and Baamba peoples. The Bakonjo have lived in the region for many generations, their culture adapted to its steep slopes and climate. Rwenzori Mountains National Park was inscribed on the list of UNESCO World Heritage Sites in 1994. Despite several interventions to interest neighbouring communities in the conservation of the Rwenzori Mountains, local communities did not show support for the park's conservation efforts (Muhmuza et al., 2009). In 2005, FFI undertook consultations about the meaning of the Rwenzori Mountains to local people. The economic and political realities of the area made interest in conservation difficult for local communities. In particular, the restrictive policies of the protected area designation were having negative impacts on local people, as they interrupted access to commune with their gods. The purely scientific values represented by protected areas created difficulties for the local communities who wanted to practice sacred rituals, gather herbs to prevent illness, and practice long-held beliefs and traditions (Muhumuza et al., 2009).

Staff from FFI and the Uganda Wildlife Authority (UWA) in 2009 undertook a consultative process among the local communities. During the consultations, the meaning and the important role of the mountain to local communities became clearer. Residents showed staff their sacred sites through which they commune with their gods up in the mountains and described their hierarchy based on ridges through which they interact and are governed. Local cultural institutions had been neglected by the central government, but the communities continued to use them and revere them. With help of the ridge leaders, 15 sacred sites were identified and geo-referenced in the Park. Through a rapid ethnographic assessment on the sacred sites from July to September 2009, staff were able to better understand the associated beliefs and importance to surrounding communities, to incorporate local beliefs into planning and to gain more acceptance and participation by surrounding communities.

The assessment showed that the sacred sites play an important role in people's daily lives. There are rituals for health, ridge cleansing, rainfall and peace. The older members of the community were both knowledgeable about the sacred sites and interested in practicing cultural rituals when permitted. Local communities believe that improper ownership and use will result in severe punishment, including death.

Staff and communities identified and mapped Rwenzori's sacred sites. The cultural research also identified strong species-human linkages through taboos and totems. The Rwenzori Cultural Association was established to champion cultural values. The park's Management Plan was then reviewed and modified to reflect cultural values, and site-level plans were developed. The park managers and community members agreed on allowable community access to sites for specific functions in the park.

5. Reasons for successes and current challenges

The aforementioned successes may be attributed to several possible factors. First, conserving life on earth is a noble cause that can provide fulfilment and self realization to individuals pursuing it. Fulfilment in this pursuit presupposes existing values sympathetic to conservation, and many of these values are religious in origin, e.g., that creation is good and should be cared for. Second, longstanding religious leadership and charisma backed up by inspiration from sacred texts can be factors in fostering conservation action by religious devotees. Third, people have the natural inclination to contribute to a noteworthy cause provided it is of social significance and confers prestige, and religion helps to establish what is significant and prestigious in many communities. People do not act as individuals, but as part of small cohorts and larger communities. When the community or cohort leaders make certain behaviours a priority, others follow (Diani, 2003; Oliver & Myers, 2003). People also become and stay involved because they want to belong and maintain their relationships with others (Aminzade & Perry, 2001; Staggenborg, 2011). Where religious institutions are strong or central they define an important community.

Mobilizing communities for conservation action based on religious beliefs and belonging to religious communities does not occur without investment in organizing. Furthermore, several obstacles must be overcome. Bhagwat et al. (2011b) note three obstacles to making religion supportive of conservation: (1) difference in worldviews (i.e., religious versus secular groups and also between religions), (2) conflict between identities (i.e., strong religious identities; this may hinder the success of relaying conservation messages to a mixed audience of different faiths) and (3) divergent attitudes and behaviour. Jacobs and The Blessed Kateri Tekakwitha Conservation Centre (2011), for example, criticizes the Earth Charter – the 1992 UN declaration of fundamental principles for building a just, sustainable and peaceful global society in the 21st Century. They write that "superficially, the Charter appears to be a noble concept designed to end social and environmental tensions around the world", but they are concerned that it furthers creation of a global super-state, which they do not support. They also note that neither Pope John Paul II nor Pope Benedict XVI endorsed the Earth Charter. This conflict in perspectives between religion and conservation may be largely due to divergent worldviews since worldviews have a strong influence on behaviour and attitudes and give particular meaning to diverse situations and events.

Although in many cases cultural values co-exist with religion in Africa, for example, especially indigenous religion, it would be instructive to know if cultural behaviour could respond to mainstream religious influence (e.g. Islam and Christianity) that stems from faith-based conservation messages. Cultural systems are often religion-based even though a religion may span many cultures, but not all elements of a culture are religiously based or justified. For instance, bushmeat consumption and trade are strong elements in many African cultures. Could such elements respond to mainstream religious influence? This calls for empirical analysis by conservation biologists and social scientists. Also, poverty is a social issue that pervades African society. Poverty pushes people to put pressure on biodiversity. Could religious values positively influence poverty-related natural resource challenges as they relate to biodiversity loss?

6. Conclusion

Based on our review, there is reason to be encouraged by this sampling of cooperative efforts of religious communities and institutions and conservationists. In the last two and a half decades since the 1986 Assisi meeting, cooperation has increased and the several joint projects undertaken have raised awareness, slowed illegal logging and poaching and found ways to benefit local communities in protecting biodiversity. If such trends are to continue, the world's religions should increase their concern and action on behalf of conservation. Likewise, the conservation community should develop a better understanding of religious values, and how they interface with conservation values.

To help bridge religion and conservation, the conservation community should reach out more effectively to religious leaders and gain a better understanding of religious concerns regarding the environment and biodiversity.

In addition to conducting research to benefit biodiversity, conservation biologists can help bridge religion and conservation by understanding differences in worldviews and by working with religious leaders and organizations to earn their trust and confidence. The linking of religion with conservation calls for direct relationships supported by conflict resolution skills, self awareness, trust and wisdom. Importantly, when conservation biologists and secular organizations interact with religious bodies, there must be a compromise of divergent perspectives so as to enable the emphasis to be on finding common grounds (e.g., saving creation), mutual respect and empathy. Indeed, in the light of this review, we see the inroads that have been made cannot be ignored. What the future holds for religion and conservation could be an upward spiral where there are an increasing number of people all over the world adopting the conservation ethic based on religious values; therefore what we need is to further explore the possibilities in the bridging of religion and conservation, as the global urgency before us leaves us little choice.

7. Acknowledgment

The authors acknowledge and thank David Johns for carefully reading and giving helpful comments to this manuscript.

8. References

Aminzade, R.R. & Perry, E.J. (2001). The Sacred, Religious and Secular in Contentious Politics. In: *Silence and Voice in the Study of Contentious Politics*, Aminzade, R.R., Goldstone, J.A., McAdam, D., Perry, E.J., Sewell, W.H., Tarrow, S. & Tilly, C.(Eds.)., pp.155-78 Cambridge University Press, Cambridge UK

Alliance of Religions and Conservation (ARC). (2011a). The Alliance of Religions and Conservation, 19.08.2011, Available from http://www.arcworld.org/about_ARC.asp

Alliance of Religions and Conservation (ARC). (2011b). ARC-UN: Faiths' Long Term Commitments for a Living Planet, 19.08.2011, Available from http://www.arcworld.org/projects.asp?projectID=358

Awolalu, J.O. (1976). What is African Traditional Religion? *Studies in Comparative Religion*, Vol.10, No.2, 24.09.2011 Available from www.studiesincomparativereligion.com

Babalola, F.D. (2009). Conservation Education: Roles of Indigenous Knowledge and Cultural Beliefs in Southwest Nigeria. *Proceedings of the International Colloquium: The Educational Research on Policy and Practice in Africa*, Bamako, Mali, December 2009

Barcott, B. (2001). For God So Loved the World. Outside Magazine Vol.3, No.26, pp.84-126

Bhagwat, S.A.; Dudley,N. & Harrop, S.R. (2011a). Religious Following in Biodiversity Hotspots: Challenges and Opportunities for Conservation and Development, *Conservation Letters*, Vol.4, (February 2011), pp. 234-240

Bhagwat, S.A.; Ormsby, A.A. & Rutte, C. (2011b). The Role of Religion in Linking Conservation and Development: Challenges and Opportunities, *Journal for the Study of Religion, Nature and Culture*, Vol.5, No.1, pp. 39-60, ISSN 1743-1689

Bhagwat, S. A. & Palmer, M. (2009). Conservation: The World's Religions Can Help, *Nature* Vol.461, No. 37, (September 2009), pp.37

Butchart, S.H.M., Walpole M., Collen B. et al. (2010). Global Biodiversity: Indicators of Recent Declines, *Science*, Vol. 328, No. 5982, (May 2010) pp.1164-1168 doi: 10.1126/science.1187512

Chen, J. (2003). *Across the Apocalypse on Horseback: Imperfect Legal Responses to Biodiversity Loss. The Jurisdynamics of Environmental Protection: Change and the Pragmatic.* Environmental Law Institute. p. 197. ISBN 1585760714

Conservation International (CI). (2005). *Islamic Boarding Schools and Conservation: The World Bank Faith and Environment Initiative* Agreement No.7133121. Final Report

Diani, M. (2003). Introduction, In: *Social Movements and Networks*, Diani, M. & Doug, M. (Eds.)., pp. 1-18 Oxford University Press, New York

Dudley, N.; Higgins-Zogib, L. & Mansourian, S. (2005). *Beyond Belief: Linking Faiths and Protected Areas to Support Biodiversity Conservation*, WWF, Retrieved from http://tinyurl.com/3ffkofy

Food and Agriculture organization of the United Nations (FAO). (2006). Global Forest Resource Assessment 2005, 18.08.2011, Available from: http://www.fao.org/docrep/006/ab780e/AB780E07.htm

Forum on Religion and Ecology (FORE). (2011). About us, 13.09.2011, Available from http://fore.research.yale.edu/about-us/

Gambrill, A. (2011). *From Practice to Policy to Practice: Connecting Faith and Conservation in Africa*, International Resources Group for USAID Bureau for Africa. Washington, DC. Available from http://www.rmportal.net/library/content/from-practice-to-policy-to-practice-connecting-faith-and-conservation-in-africa/

Hall, M.; Grim, J. & Tucker, M.E. (2009). Need for Religions to Promote Values of Conservation, *Nature*, Vol.462, No.10, (December 2009), pp.720

Higgins, S. (2011). Conservation with Heart: Bridging Science and Religion, 19.08.2011, Available from http://tinyurl.com/6xww37y

International Union for the Conservation of Nature (IUCN). (2011). About the Biodiversity Crisis, 15.08.2011, Available from: http://www.iucn.org/what/tpas/biodiversity/

Jacobs, B. & Blessed Kateri Tekakwitha Conservation Centre (2011). The Earth Charter: Constitution of the Global Super-State? A Catholic Reflection on the Earth Charter, 13.09.2011, Available from http://conservation.catholic.org/Earth%20Charter.htm

Kellert, S.R. and T. Farnham (2002). *The Good in Nature and Humanity: Connecting Science, Religion, and Spirituality with the Natural World*, Washington, DC: Island Press

Millennium Ecosystem Assessment (2005). *Ecosystems and Human Well-Being: Biodiversity Synthesis*, World resources Institute, Island Press, Washington D.C, Retrieved from http://www.millenniumassessment.org/documents/document.354.aspx.pdf

Muhumuza, M.; Biira, O.; and Mugisha, A. (2009). *The Role of Cultural Values in the Conservation of the Rwenzori Mountains National Park.* presented at Nature Uganda conference, 19-20 November, 2009. Published in conference proceedings and submitted for publication in the *Journal of East African Natural History.* Available from http://www.mmu.ac.ug/publications/96-muhumuza-moses.html#mm5

Murray, J. (May 2010). UN-Biodiversity Crisis Threatens Global Economy, 19.08.2011, Available from http://tinyurl.com/6k2bsxe

Novacek, M.J. (2008). Engaging the Public in Biodiversity Issues, 18.08.2011, Available from www.pnas.org/cgl/dol/10.1073/pnas.0802599105

Oliver, P.R. & Myers, D. (2003). Networks, Diffusion and Cycles of Collective Action. In: *Social Movements and Networks*, Diani, M. & Doug, M. (Eds.)., pp.173-203 Oxford University Press, New York

One Country. (1995). Religions Vow a New Alliance for Conservation, 19.08.2011, Available from http://www.onecountry.org/oc71/oc7101as.html

Otabil, A. (2010). Religious Bodies Partner Government on Climate Change, 25.10.2011, Available from http://tinyurl.com/69zbaj6

Rappaport, R.A. (1974). "Sanctity and Adaptation". Coevolution Quarterly No. 2 (Summer), pp.54-68, The article is based on a presentation to the Wenner-GrenFoundation Conference "The Moral and Esthetic Structure of HumanAdaptation", 1969

Rappaport, R.A. (1976). Adaptations and Maladaptations in Social Systems, In: *The Ethical Basis of Economic Freedom*, Hill, I. (Ed.)., pp.39-79, Chapel Hill, NC: American Viewpoint

Rappaport, R. A. (1979). *Ecology, Meaning and Religion.* 2nd edition. North Atlantic Books, Berkeley, California

Rappaport, R. A. (1999). *Ritual and Religion in the Making of Humanity,* Cambridge University Press, Cambridge, United Kingdom

Rolston III H. (2010). Saving Creation: Faith Shaping Environmental Policy, *Harvard Law & Policy Review* Vol.4, pp.121-148

Sigmar, G. (2007). Biodiversity 'Fundamental' to Economics, 13.09.2011, Available from http://tinyurl.com/3n52exs

Sluka, R.D.; Kaonga, M.; Weatherly, J.; Anand, V.; Bosu, D. & Jackson, C. (2011). Christians, Biodiversity Conservation and Poverty Alleviation: a Potential Synergy? *Biodiversity*, Vol.12, No.2, pp.108-115

Sponsel, L.E. (2007). Religion, Nature and Environmentalism, In: *Encyclopaedia of Earth*, Eds. Cutler J.C. (Washington, D.C. Environmental Information Coalition, National Council for Science and the Environment), 09.07.2011, Available from http://www.eoearth.org/article/Religion,_nature_and_environmentalism

Staggenborg, S. (2011). *Social Movements*, Oxford University Press, New York

The International Small Group and Tree Planting Program (TIST). (2012). www.tist.org

United Nations Educational, Scientific and Cultural Organization (UNESCO). (2011). Osun Osgbo Sacred Grove, 13.09.2011, Available from http://whc.unesco.org/en/list/1118

Verschuuren, B., R. Wild, J. McNeely, & G. Oviedo. (2010) *Sacred natural sites: Conserving nature and culture*. London: Earthscan.

Wilkinson, C. (Ed.). (2008) *Status of Coral Reefs of the World: 2008*. Global Coral Reef Monitoring Network and Reef and Rainforest Research center, Townsville, Australia. Retrieved from http://unesdoc. unesco.org/images/0017/001792/179244e.pdf

Wilson, D.S. (2002). *Darwin's Cathedral: Evolution, Religion, and the Nature of Society*. University of Chicago Press, Chicago

World Bank. (2006). *Faiths and the Environment World Bank Support 2000 – 2005*, Retrieved from http://tinyurl.com/ylp3bn

Yale University. (2011). Science, Religion and the Natural World Theme of Conference at Yale May 11-14, 09.10.2011, Available from http://opac.yale.edu/news/article.aspx?id=5436

6

Managing Population Sex Ratios in Conservation Practice: How and Why?

Claus Wedekind

*Department of Ecology and Evolution, Biophore,
University of Lausanne, Lausanne,
Switzerland*

1. Introduction

Small or declining populations are at increased risk of extinction because of stochasticity and Allee effects (Lande 1998, Courchamp et al. 1999, Stephens and Sutherland 1999, Bourbeau-Lemieux et al. 2011), and several genetic problems that include reduction in genetic variability, an accumulation of deleterious mutations due to random drift, and increased rates of inbreeding depression (Frankham et al. 2002, Hedrick 2005, Allendorf and Luikard 2007). Genetic problems are likely to reduce the average viability of individuals from generation to generation, and they reduce evolutionary potential and therefore the long-term survival expectancies, especially of small populations (Frankham et al. 2002, Hedrick 2005, Allendorf and Luikard 2007). However, genetic problems are only indirectly linked to the census size (N_c). Instead, they are directly dependent on the genetically effective population size (N_e) that is defined as the size of an ideal model population that looses genetic variability at the same rate as the observed population. Usually, N_e is significantly smaller than N_c because of variance in individual reproductive success, deviations from a 1:1 operational sex ratio, and other reasons. Risks of extinction are therefore increased if population sex ratios deviate from 1:1.

We typically expect 1:1 sex ratios in natural populations because of strong frequency-depended selection on the production of sons and daughters (Fisher 1930). However, population sex ratios can be biased by non-random harvest as a consequence of, for example, sex differences in behavior, size, or morphology, or simply as a consequence of hunter preferences (Bunnefeld et al. 2009, Tryjanowski et al. 2009, Marealle et al. 2010). Sex ratios can also be influenced by environmental changes such as, for example, different kinds of chemical pollution or changes in the temperature regime that may cause sex-specific mortality or growth. Environmental changes can even directly influence the production of males and females in species with environmental sex determination (Janzen 1994, Kamel and Mrosovsky 2006), or in species where the genetically determined sex can be reversed during a critical period in life. Such environmental sex reversal has been observed in several fish and amphibians (Wallace et al. 1999, Devlin and Nagahama 2002, Baroiller et al. 2009, Stelkens and Wedekind 2010), may potentially be more likely under many of the rapid environmental changes we are currently observing, but may well have happened frequently even before anthropogenic effects on the environment became ubiquitous (Perrin 2009).

Lastly, parents (especially mothers) of many species are able to manipulate family sex ratio, as will be explained below. There are examples where the combined effects of such parental life-history decisions have lead to distorted population sex ratios (Robertson et al. 2006).

We may be able to manipulate and hence manage population sex ratios to benefit biodiversity if we understand how they are influenced under natural and artificial conditions. We may either aim for maximizing the evolutionary potential and hence the long-term perspectives of a given population, or wish to control the growth of problem populations (e.g. of exotic species). Among the various tools that have been proposed for manipulating sex ratios are the 'sterile male' strategy, the 'Trojan Y chromosome', and recombinant constructs that lead to gender distortion (Gutierrez and Teem 2006, Cotton and Wedekind 2007a, Bax and Thresher 2009). Alternatively, maternal life-history strategies can sometimes be manipulated in order to affect family sex ratios, and some species even allow for sex ratio manipulation by simple manipulations of the micro-ecological conditions during critical stages in ontogeny.

In the following I summarize the current knowledge about how population sex ratios develop, and how they can change due to, for example, changed temperature regimes, different kinds of chemical pollution, or other environmental changes. I will then outline the various tools that could be used to manipulate sex ratios and give some examples from the literature. I will discuss the potential risks and benefits of such manipulations, and I will list a number of key questions that still need to be answered in order to optimize the management of population sex ratios.

2. What affects family sex ratios?

When discussing family sex ratios, it is useful to distinguish between the different possible explanatory levels, especially between proximate and ultimate explanations (Tinbergen 1963). Proximate (mechanistic) explanations of family sex ratio deal with questions about the genetic, physiological, and molecular aspects of, for example, sex determination. Ultimate (evolutionary) explanations concentrate on the adaptive value of a given family sex ratio, especially on the impact of a parent's fitness, without necessarily explaining the proximate aspects. Obviously, proximate arguments often set constraints to what parents may be able to achieve in order to maximize fitness.

If sex determination is purely environmental, as in most reptiles, sex is not determined at conception but later during a specific window of time during embryonic or larval development. The window is often called "the thermosensitive period" because incubation temperature is often the most important sex-determining factor in these species (Valenzuela and Lance 2004). Purely environmental sex determination has been assumed to be quite common also in fish. However, Ospina-Alvarez and Piferrer (2008) argued that among the many species for which sex-determining chromosome have not (yet) been identified, species should only be considered as having a purely environmental sex determination if sex is determined by environmental conditions that can be considered as normal and within the range usually experienced under natural conditions. Applying this condition leaves only few species of four teleost orders with purely environmental sex determination. Among them, three different types of reaction norms dominate: (i) decreased or (ii) increased frequency of males with increasing temperature, or (iii) high frequency of males at extreme (high or low) temperatures (Devlin and Nagahama 2002, Ospina-Alvarez and Piferrer 2008, Baroiller et al. 2009).

In species with environmental sex determination, the within-population variance in family sex ratio can be very high due to variance in the micro-ecological conditions that affect eggs or larvae. Moreover, regional changes in the environment can easily lead to skewed population sex ratios in some years (Janzen 1994, Kamel and Mrosovsky 2006). Rapid and consistent environmental changes could then have dramatic consequences on population growth especially in small population of limited genetic variability, or in fragmented populations with limited gene flow. However, correlated changes in nesting or spawning time (Janzen et al. 2006, Wedekind and Küng 2010) or other changes in behavior and life history could potentially mitigate some of these negative effects of environmental changes. Moreover, some species seem to have the potential for (rapid) evolution in response to changed environments (Conover and Van Voorhees 1990, Conover et al. 1992, Magerhans et al. 2009).

In many fish and amphibians, sex determination is genetic but reversible by environmental factors during a sensitive period that is typically very early in life. Environmental sex reversal can be induced by various factors, including temperature changes or exposure to hormone active substances (Wallace et al. 1999, Devlin and Nagahama 2002, Baroiller et al. 2009). It is nowadays even used in fish farming to produce more profitable one-sex cultures (Pandian and Sheela 1995, Piferrer 2001, Cnaani and Levavi-Sivan 2009). Distorted sex ratios in the wild could potentially be caused by environmental sex reversal (Olsen et al. 2006, Brykov et al. 2008, Alho et al. 2010). Sex hormones, hormone-active substances, and endocrine disrupting chemicals are frequently released into natural watercourses, for example, in effluents from domestic and industrial sources (Larsson et al. 2000, Parks et al. 2001, Jobling and Tyler 2003). Fish exposed to such chemicals often display reduced reproductive performance (Vos et al. 2000), and exposure to such chemicals could well be responsible for gonadal malformations if, for example, sex reversal was incomplete leading to individuals that display gonadal characteristics of both sexes. A sudden increase in the prevalence of intersex or of other gonadal malformations is indeed frequently observed in natural populations (Harries et al. 1997, Bernet et al. 2004, Penáz et al. 2005, Jobling et al. 2006, Bernet et al. 2008, Bittner et al. 2009). Other possible consequences of exposure to hormones or hormone-active substances may include reductions in gonadal growth, a delayed onset of sexual maturity, inhibition of spermatogenesis, lower egg production, or reduced egg quality (Sumpter and Jobling 1995) (Vos et al. 2000). However, sex ratios in the wild can be skewed for many reasons (Palmer 2000), and environmentally induced sex reversal is often difficult to prove (Nagler et al. 2001, Chowen and Nagler 2004, 2005, Williamson et al. 2008). The prevalence and significance of environmental sex reversal in the wild is therefore still unclear (Wedekind 2010). So far, the consequences of environmentally induced sex reversal have only been analyzed in theoretical studies (Kanaiwa and Harada 2002, Hurley et al. 2004, Cotton and Wedekind 2009). These studies suggest that environmentally induced sex reversal can change population growth and population sex ratios in ways that may sometimes be counter-intuitive. A moderate rate of feminization, i.e. of an environmentally-induced development of the female phenotype despite male sex chromosomes, could sometimes be beneficial for population growth, especially in the absence of strong viability effects of the sex reversal. However, most possible outcomes of environmental sex reversal are negative with regards to population growth or the persistence of sex chromosomes. For example, strong environmental feminization over several generations leads to high rates of YY individuals and can eventually lead to the extinction of X chromosomes (Cotton and Wedekind 2009). Analogously, continuous environmental masculinization increases the rate of XX individuals and can drive the Y chromosome to extinction (Cotton and Wedekind 2009). If sex chromosomes are lost, i.e. if populations loose their genetic sex determination in response to environmental factors that induce sex reversal, the affected population may quickly go extinct if the environmental forces that cause sex reversal cease.

The frequency-dependent selection on the production of sons and daughters is a consequence of the fact that every sexually produced individual usually has exactly one father and one mother. This explains why 1:1 sex ratios are so common. Such equal sex ratios are easier to achieve if sex determination in purely genetic as compared to if sex determination is environmentally biased. Sex determination is purely genetic in all mammals and birds and in many species of other taxa. However, even in these taxa, equal primary sex ratios are a rule with exceptions, and parents of many species are often able to somehow manipulate family sex ratio. The physiology of such manipulations is often not clear yet, but there are good reasons why skewed family sex ratio may offer fitness benefits. If, for example, one sex is more costly to produce and raise than the other one, parents who are able to weight the relative investment into sons and daughters according to the expected fitness return would achieve higher fitness than parents who would not be able to do so (Charnov 1982).

The relative value of sons and daughters may differ for different parents. This is especially so if the expected fitness return of one type of offspring is more dependent on resources received from the parents than the expected fitness return of the other type of offspring. This is a typical outcome in polygamous species where, for example, few dominant males receive an increased reproductive success at the expense of outcompeted males. In such cases, parents that are in good condition or experience favorable conditions would increase their fitness by producing more offspring of the sex with the greater requirements, while parents in suboptimal condition should rather "play safe" and produce the other sex that promises a lower variance in reproductive success (Trivers and Willard 1973). This hypothesis received much empirical support in a variety of taxa (Gomendio et al. 1990, Cassinello and Gomendio 1996, Bradbury and Blakey 1998).

If family sex ratio can be adjusted in response to parental condition, it seems reasonable to assume that mate attractiveness could influence overall parental investment and maybe even family sex ratio. Indeed, females of various taxa have been observed to produce relatively more sons if mated with an attractive male rather than if mated with an less attractive male (Burley 1982, Ellegren et al. 1996), possibly because sons of attractive males may generally be more likely to achieve high reproductive success than sons of less attractive males. However, the effect could not always be observed (Westerdahl et al. 1997, Saino et al. 1999). Moreover, skewed family sex ratios do not necessarily reflect adaptive parental strategies (Krackow 1995).

3. Managing population sex ratio

3.1 Why?

There are several good reasons why conservation managers could potentially profit from manipulating population sex ratios (Table 1). First, sex ratios of a small and endangered population are sometimes significantly skewed towards too many females or too many males. Such non-equal sex ratios increase the negative effects of Allee effects and demographic stochasticity, sometimes leading to extinction. For example, the last six individuals of the Dusky Seaside Sparrow (*Ammodramus maritimus nigrescens*) that remained and were kept in a captive breeding program turned out to be all male. This subspecies is now believed to be extinct (en.wikipedia.org from Feb 6th 2012). Non-equal sex ratios also reduce the genetically effective population size because $N_e = 4N_mN_f/(N_m+N_f)$, with N_m and N_f being the number of mature males and females, respectively (Hartl 1988). In harem-based mating systems, N_e is even better described as $N_e = 4N_mN_f/(2N_m+N_f)$ (Nomura 2002). Therefore, populations with non-equal sex ratios are expected to lose more genetic variability (i.e. evolutionary potential) and suffer more from the negative consequences of inbreeding than populations with equal sex ratios. Moreover, in species where population growth is limited by the availability of oocytes, male-biased sex ratios directly reduce population growth. For example, the remaining population of the critically endangered kakapo (*Strigops habroptilus*) of New Zealand has been found to be significantly male biased, probably as an undesirable side effect of supplementary feeding (Tella 2001, Clout et al. 2002). Because this parrot is also a species for which population growth is severely limited by egg production, manipulating family sex ratios towards more daughters (by methods that are outlined below) quickly became one of the priorities of the conservation management of this species (Robertson et al. 2006).

Second, if population sex ratios are not significantly skewed, but population size is small and population growth is limited by the availability of oocytes, manipulating sex ratio towards a female bias could sometimes be desirable, especially if very small or rapidly declining populations call for emergency actions. For example, captive breeding programs are typically not only meant as a refuge in response to a temporary ecological crisis, but they often aim at releasing additional individuals into the wild to support small or declining populations and to help preventing further losses of genetic diversity (Young and Clarke 2000). Such captive breeding programs sometimes even include the use of assisted reproductive technologies (Gibbons et al. 1995, Dobson and Lyles 2000, Lanza et al. 2000). There are a number of potential risks that need to be considered in such programs. These risks include, among others, a general increase in the variance in reproductive success and hence a reduction in overall N_e (Ryman and Laikre 1991), potential negative effects of circumventing natural mate preferences (Grahn et al. 1998, Wedekind et al. 2001, Wedekind 2002b, Jacob et al. 2010), and artificial selection favoring certain life-history characteristics (Heath et al. 2003, Wedekind et al. 2007). However, if we can assume that offspring have an enhanced survival in captivity and that the subsequent release of captive bred individuals into the wild has a positive effect of the long-term survival of the population, artificially changing family sex ratios towards a female bias could sometimes even further increase population growth rate. Such manipulations may be feasible within a captive breeding program, for example by manipulating female reproductive strategies, or directly within the

Observation	Aim of intervention	Main reasons for intervention**	Potential intervention technique	Immediate effect of the intervention	Typical taxa for which the intervention may be most promising	Typical taxa for which the intervention is not likely to work
Biased sex ratio in small population, not caused by ESR*	Remove sex ratio bias	1, 2, 3, 4	Manipulate parental strategies	Changes sex ratio of next generation	Polygamous species, or monogamous species with high rates of extra-pair fertilizations	Species with equal reproductive potential for males and females
			Manipulate environment during embryogenesis or early larval stages	Changes sex ratio of next generation	Species with environmental sex determination (e.g. turtles, crocodiles, some fishes)	Species with genetic sex determination and no ESR* (e.g. most mammals and birds)
The growth of a small population is limited by the availability of oocytes	Increase population growth rate	1	Manipulate parental strategies	Changes family sex ratio in order to create a female excess	Polygamous species, or monogamous species with high rates of extra-pair fertilizations	Species with equal reproductive potential for males and females
			Manipulate environment during embryogenesis or early larval stages	Changes sex ratio of next generation to create a female excess	Species with environmental sex determination	Species with genetic sex determination and no ESR
			Introduction of Trojan sex chromosome carriers	Daughter-biased family sex ratio in introduced individuals	Fishes and amphibians with ZZ/ZW sex determination system, ESR, and functional W-chromosome	Species without ESR; species with XX/XY sex determination system; species with decayed W-chromosome
Biased sex ratio caused by ESR*	Avoid short-term and long-term consequences of ESR; increase population growth	1, 2, 3, 4, 5	Identify and control driver of ESR	Genotype-phenotype mismatch reduced; strong immediate effects on population size and sex ratio possible (caution : continuous ESR could lead to extinction of one of the sex chromosomes)	Many fishes and amphibians	Species with strict environmental or genetic sex determination
Damaging effects of invasive species	Control or reduce population growth rates	6	Introduction of 'Trojan Y chromosomes'	Male-biased family sex ratio in introduced individuals	Many fishes and amphibians with XX/XY sex determining system and functional Y chromosome	Species with strict genetic sex determination or ZZ/ZW sex determination system; species with decayed Y chromosomes
			Introduction of sterile males	Immediate male bias, increased male-male competition, increased number of non-fertilized eggs	Many species	?
			Introduction of the recombinants "daughterless"	Male-biased family sex ratio in introduced individuals	Some fishes	All other species

* Environmentally-induced sex reversal, i.e. mismatch between gender genotype and phenotype
** 1. Typical problems of small populations, i.e. increased risk of extinction because of demographic and environmental stochasticity, Allee effects, or various genetic problems
2. Increased two-sex demographic stochasticity (i.e. increased risk of loosing one sex)
3. Reduced N_e/N_c ratio (i.e. increased genetic drift, reduction of overall heterozygosity, inbreeding depression)
4. Increased risk of losing cultural traits, especially if the sex ratio bias or the low population size is untypical
5. Increased risk of extinction of a sex chromosome
6. Various possible negative effects of an invasion, including a reducing biodiversity, altering ecosystem processes, causing economic losses, exotics acting as vectors of new diseases, etc.

Table 1. Circumstances where different forms of sex ratio manipulation may or may not be warranted (see text for further explanations and for references).

wild population, for example by releasing individuals that are more likely to produce daughters than sons (Cotton and Wedekind 2007b) (see below). It is important to note that such a sex-ratio manipulation has the immediate effect of reducing the N_e to N_c ratio (because $N_e = 4N_mN_f/(N_m+N_f)$), i.e. it increases the genetic bottleneck that the small population is going through. This increased bottleneck immediately increases demographic stochasticity and leads to a greater loss of genetic variance, higher inbreeding rates, and higher rates of genetic drift and hence of fixation of deleterious mutations. However, if the sex ratio manipulation is carefully applied, these immediate negative effects of the treatment can be outweighed by the increased reproduction rates and the accelerated population growth (Wedekind 2002a, Lenz et al. 2007, Cotton and Wedekind 2009).

Third, invasions by exotic species, for example after a planned or accidental release of a non-native species, are a major threat to biodiversity in most regions of the world (Myers et al. 2000). Various methods have been proposed to deal with this threat, but some of these methods have frequently created further problems, e.g. the introduction of secondary controlling species (Louda and Stiling 2004). Biasing the sex ratio in such problem populations towards more males could be a largely reversible method that may not only reduce population growth but could even reduce average female fitness. Male harassment of females over mating could by itself accelerate population decline (Rankin and Kokko 2007). At very low population sizes, induced male biases could even enhance Allee effects (Stephens and Sutherland 1999). Sex ratio manipulation may hence be an attractive option in fighting or controlling exotic species.

Family sex ratio manipulation is possible in many taxa. The degree of the invasiveness of the manipulation spans from manipulating environmental conditions during embryo and larval development or manipulating female perception of certain environmental key factors to, for example, sperm sexing prior to assisted reproductive technology in captive breeding programs (Gibbons et al. 1995, Dobson and Lyles 2000, Lanza et al. 2000). With regard to the latter, some methods of micromanipulation and some *in vitro* culture conditions have been discussed as potentially having an effect on embryo sex ratio in mice and cattle (King et al. 1992, Gutierrez et al. 1995). It therefore seems possible that sex ratios could be manipulated if assisted reproductive technology is used to propagate a species. However, in the following discussion of sex ratio manipulation, I will concentrate on methods that are arguably less invasive and comparatively less expensive.

3.2 Manipulating the rearing environment or maternal decisions

Obviously, if sex determination is purely environmental, a simple manipulation of the environment that embryos, larvae, or juveniles experience during the critical window in time in which sex is determined can be sufficient. If this critical time is during egg development, as for example in most if not all turtles, eggs could be collected and incubated at temperatures that result in the desired family sex ratio. Alternatively, the conditions at the egg laying site could be artificially changed (Girondot et al. 1998). Analogous manipulations have been suggested from some amphibians (Solari 1994).

Manipulating family sex ratio is less straightforward in species with genetic sex determination. However, the frequent observation that females are somehow able to manipulate family sex ratio (or the sex of their one offspring) in response to ecological or social characteristics of the rearing environment may provide a number of options. If, for example, females adjust their family sex ratio in response to a perceived skew in the population sex ratio, skewed sex ratios could potentially be simulated in captive populations, for example by removing and housing members of one sex separately. Alternatively, the sensory stimuli that females use to perceive their social environment could be manipulated, for example by exposing the female to urine of different individuals in order to simulate a skewed population sex ratio (Perret 1996).

The Trivers-Willard hypothesis (Trivers and Willard 1973) predicts that in polygynous species, females in good conditions are more likely to have sons than daughters (see above). Such parental decisions could potentially be manipulated by manipulating the females' condition, for example by a changed feeding regime. The kakapo may be an example here. Supplementary feeding of the few remaining individuals of that species may have led to male-biased sex ratios because females in good conditions turned out to be more likely to have sons than daughters (Tella 2001, Clout et al. 2002). At one point in time, about 70% of all recorded offspring of this species were sons. Robertson et al. (2006) found that the male bias was significantly reduced when female condition was altered. Lenz et al. (2007) used this line of thought to work out the likely genetic and demographic consequences of analogous management actions in an existing captive breeding program for a Spanish population of the lesser kestrel (*Falco naumanni*), another polygynous bird that shows a correlation between family sex ratio and female condition: more daughters are born by mothers of average conditions, while more son are born by mothers of good condition (Aparicio and Cordero 2001). The authors found that a sex-ratio management within the range that seems possible would significantly increase the efficiency of an existing captive breeding program.

If females adjust their investment into sons and daughters according to male characteristics, it may be possible to exploit the rules that females use to determine the attractiveness of a male relative to all potential mates. Such decision rules are not likely to be entirely genetically fixed but may be rather flexible (Real 1991). Mate choice decisions can be the outcome of simple cost/benefit analyses (Milinski and Bakker 1992), and the perception of the attractiveness of a given male is expected to depend on female experience and hence on a sampling template given by the population. Such a sampling template could be manipulated in order to increase or decrease the perceived attractiveness of a given male. If, for example, the size or the color of a secondary sexual ornament determines sexual attractiveness, exposing the female to several (real or dummy) individuals with very weak sexual ornaments may make a male with medium-sized or medium–colored ornament more attractive. Analogously, exposing the female to several individuals with strong sexual ornaments may make the male with a medium-sized or a medium-colored ornament less attractive. Alternatively, many secondary sexual ornaments could be directly manipulated. Structural ornaments could be artificially elongated, colors could be enhanced painting the ornament, or the male could be presented under light conditions that accentuate the colors in question.

3.3 'Trojan Y chromosomes' and genetic constructs that distort sex ratios

Conservation practice sometimes includes managing potential problem populations (Kolar and Lodge 2002, Hanfling et al. 2011, Poulin et al. 2011). The 'sterile male' strategy is one of the various techniques that has been proposed. The idea is that large numbers of sterile males are produced and released into the wild in order to outcompete wild males for mating. So far, the application of this idea has largely concentrated on disease-transmitting insects (Thailayil et al. 2011). However, if females mate with only few males each, and the mating frequency of introduced sterile males is not significantly smaller than the mating frequency of wild males, the 'sterile male' strategy could potentially be applied more widely and also in the context of exotic species that need to be controlled. And even if the idea behind this technique is not primarily based on changing population sex ratio in order to manipulate population growth, it should nevertheless be discussed in the present context. Obviously, releasing large numbers of sterile males leads to a male-biased sex ratio that may, by itself, increase intra-individual competition and lead to a reduction of average female fitness not only because of the increased rate of non-fertilized eggs but also because of the likely negative effects of male-biased sex ratios in some sort of species (Rankin and Kokko 2007). Further recombination methods that have been discussed as possible pest control include sex-specific lethality constructs (Schliekelman and Gould 2000, Schliekelman et al. 2005). The effectiveness of the release of such constructs can be greatly enhanced by complementary management options such as selective harvest of males or females (Bax and Thresher 2009). However, the recombinant approach could lead to undesirable results if, for example, the gene construct jumps to other species (Kapuscinski et al. 2007).

Species with predominantly genetic but environmentally reversible sex determination (i.e. many fish and amphibians) offer another approach, the so-called 'Trojan Y chromosomes' strategy ('Trojan' genetic elements were originally defined as elements that have the potential of driving local populations to extinction (Muir and Howard 2004)). The necessary prerequisites are that (i) the species in question displays male heterogamety (i.e. XX = female and XY = male), and that (ii) the Y-chromosome should not be significantly decayed as it usually is, for example, in most mammals because of the suppressed recombination between the sex chromosomes (Bull 1983, Rice et al. 2008) and the thereby resulting accumulation of deleterious mutations on the Y-chromosome (Muller 1932, Felsenstein 1974). Interestingly, in fish and amphibians with genetic but environmentally reversible sex determination, sex chromosomes are typically not heteromorphic, and the functionality of Y chromosomes seems mostly unrestricted. This for itself leads to a number of interesting evolutionary questions (Perrin 2009), but what is probably most interesting in the present context is the fact that YY individuals are therefore often viable. Such YY individuals can be produced by mating feminized XY individuals with wild-type XY males. YY individuals would normally be males who can only produce XY sons if mated with wild-type XX females. Sex-reversed YY individuals (i.e. females without X chromosomes) would also be expected to have only sons if mated with a wild-type XY male. Moreover, half of these sons would have the YY genotype and could hence only have sons themselves. Gutierrez and Teem (2006) modeled the repeated introduction of YY females as potential tool in conservation management. They found that such introduction of 'Trojan Y chromosomes' can potentially be used to control the growth of problem populations. Critical variables in these scenarios are the relative viability of carriers of these 'Trojan Y chromosomes' and their attractiveness in mate choice, i.e. their mating success relative to the wild-type females and males (Cotton and Wedekind 2009).

Some species that display environmentally reversible sex determination have a sex determination mechanism that is based on female heterogamety (i.e. ZZ = male and ZW = female). Introducing sex-reversed WW individuals would then lead to an increased population growth, especially if the induced sex change had no significant effect on viability and mating success (Cotton and Wedekind 2007b). Boosting population growth with 'Trojan sex chromosomes' may currently have the highest potential in the conservation of amphibians. Many amphibian display female heterogamety (Hillis and Green 1990), are susceptible to environmental influences during sex determination (Wallace et al. 1999), and WW genotypes may generally be viable and fertile because W chromosomes do typically not seem to be decayed (Perrin 2009).

The potential of 'Trojan sex chromosomes' for boosting or reducing population growth still needs to be experimentally analysed. For now, the concept seems entirely theoretical, i.e. to the best of my knowledge no empirical test of this idea has been published so far. The same seems to be true for a genetic construct that has recently been suggested and modeled by Bax and Thresher (2009) and that would induce a shift in the sex ratio of fish population. The idea here is that if individuals with multiple copies of a genetically engineered aromatase inhibitor gene (D) are introduced into a problem population, all offspring of the D gene carrier that inherit the D gene would phenotypically develop into males regardless of the composition of their sex chromosome. Analogously to Gutierrez and Teem's (2006) original 'Trojan sex chromosomes' idea, the introduction of the D gene into a population could shift the sex ratio in future generations to a male bias that potentially reduces population growth.

4. Conclusions

It may often be possible to manipulate population sex ratios, for example by changing certain ecological or social factors that influence maternal decisions about family sex ratio, or by invasive techniques like, for example, introducing sex-reversed individuals into natural populations to boost or reduce population growth on the long run. If the aim of such manipulations is to support a small and endangered population, it is important to consider the possible dangers of the manipulation. If the sex ratio of a small population is found to be male-biased prior to the intervention, reducing this bias in future generations may generally be beneficial because this would be increasing the N_e to N_c ratio and thereby reducing the negative effects of small population size on population genetics. However, even if we deal with populations in which sex ratio directly determines population growth, any deviation from equal sex ratio towards a female-biased sex ratio reduces the N_e to N_c ratio, i.e. It may create a genetic bottleneck. On the long run, the likely negative effects of such a bottleneck would need to be compensated by the increased population growth that was achieved through the sex ratio manipulation (as, for example, modeled in Lenz et al. (2007)). Furthermore, by changing a population sex ratio we are changing demographic parameters that may significantly influence breeding systems, mate choice, sex-specific use of resources, or other life-history aspects (Emlen and Oring 1977, Andersson 1994). It may even be possible that we thereby risk losing culturally transmitted characteristics that could be linked to, for example, natural breeding systems. The potential costs and benefits of a sex ratio manipulation should therefore carefully optimized for any given situation, i.e. the optimal sex ratio manipulation is likely to differ from case to case.

In the case of small and declining populations, any kind of sex ratio manipulation is likely to fail if the underlying stressors and threats to the population are not appropriately dealt with. Moreover, many of the ideas discussed here are relatively new and lack empirical support. For example, the potential of manipulating female strategies in a given species is often unclear, and we need to learn more about the viability and the fertility of sex-reversed individuals in the wild to better estimate the potential of the 'Trojan Y chromosome' strategies to control or boost populations. However, it is clear that population sex ratio can be managed in attempting to reduce genetic bottlenecks and the effects of stochasticity in small or declining populations, and to control the spread of invasive species.

5. Acknowledgments

I thank Manuel Pompini and Tony Povilitis for comments on the draft manuscript, and the Swiss National Science Foundation for financial support.

6. References

Alho, J. S., C. Matsuba, and J. Merilä. 2010. Sex reversal and primary sex ratios in the common frogs (*Rana temporaria*). Molecular Ecology 19:1763-1773.

Allendorf, F. W. and G. Luikard. 2007. Conservation and the genetics of populations. Oxford University Press, Malden, MA, USA.

Andersson, M. 1994. Sexual selection. Princeton University Press, Princeton.

Aparicio, J. M. and P. J. Cordero. 2001. The effects of the minimum threshold condition for breeding on offspring sex-ratio adjustment in the lesser kestrel. Evolution 55:1188-1197.

Baroiller, J.-F., H. D'Cotta, and E. Saillant. 2009. Environmental effects on fish sex determination and differentiation. Sexual Development 3:118-135.

Bax, N. J. and R. E. Thresher. 2009. Ecological, behavioral, and geetics factors influecing the recombinant control of invasive pests. Evolutionary Applications 19:873-888.

Bernet, D., A. Liedtke, D. Bittner, R. I. L. Eggen, S. Kipfer, C. Küng, C. R. Largiadèr, M. J. F. Suter, T. Wahli, and H. Segner. 2008. Gonadal malformations in whitefish from Lake Thun: Defining the case and evaluating the role of EDCs. Chimia 62:383-388.

Bernet, D., T. Wahli, C. Küng, and H. Segner. 2004. Frequent and unexplained gonadal abnormalities in whitefish (central alpine *Coregonus* sp.) from an alpine oligotrophic lake in Switzerland. Diseases of Aquatic Organisms 61:137-148.

Bittner, D., D. Bernet, T. Wahli, H. Segner, C. Küng, and C. R. Largiadèr. 2009. How normal is abnormal? Discrimination between deformations and natural variation in gonad morphology of European whitefish *Coregonus lavaretus*. Journal of Fish Biology 74:1594-1614.

Bourbeau-Lemieux, A., M. Festa-Bianchet, J. M. Gaillard, and F. Pelletier. 2011. Predator-driven component Allee effects in a wild ungulate. Ecology Letters 14:358-363.

Bradbury, R. R. and J. K. Blakey. 1998. Diet, maternal condition, and offspring sex ratio the zebra finch, Poephila guttata. Proceedings of the Royal Society of London Series B-Biological Sciences 265:895-899.

Brykov, V. A., A. D. Kukhlevsky, E. A. Shevlyakov, N. M. Kinas, and L. O. Zavarina. 2008. Sex ratio control in pink salmon (*Oncorhynchus gorbuscha* and chum salmon (*O-keta*) populations: The possible causes and mechanisms of changes in the sex ratio. Russian Journal of Genetics 44:786-792.

Bull, J. J. 1983. Evolution of sex determining mechanisms. Benjamin/Cummings Publishing Company, Inc, Menlo Park, CA.

Bunnefeld, N., D. Baines, D. Newborn, and E. J. Milner-Gulland. 2009. Factors affecting unintentional harvesting selectivity in a monomorphic species. Journal of Animal Ecology 78:485-492.

Burley, N. 1982. Reputed band attractiveness and sex manipulation in zebra finches. Science 215:423-424.

Cassinello, J. and M. Gomendio. 1996. Adaptive variation in litter size and sex ratio at birth in a sexually dimorphic ungulate. Proceedings of the Royal Society of London Series B-Biological Sciences 263:1461-1466.

Charnov, E. L. 1982. The theory of sex allocation. Princeton University Press, Princeton.

Chowen, T. R. and J. J. Nagler. 2004. Temporal and spatial occurrence of female Chinook salmon carrying a male-specific genetic marker in the Columbia River watershed. Environmental Biology of Fishes 69:427-432.

Chowen, T. R. and J. J. Nagler. 2005. Lack of sex specificity for growth hormone pseudogene in fall-run Chinook salmon from the Columbia River. Transactions of the American Fisheries Society 134:279-282.

Clout, M. N., G. P. Elliott, and B. C. Robertson. 2002. Effects of supplementary feeding on the offspring sex ratio of kakapo: a dilemma for the conservation of a polygynous parrot. Biological Conservation 107:13-18.

Cnaani, A. and B. Levavi-Sivan. 2009. Sexual development in fish, practical applications for aquaculture. Sexual Development 3:164-175.

Conover, D. O. and D. A. Van Voorhees. 1990. Evolution of a balanced sex ratio by frequency-dependent selection in a fish. Science 250:1556-1558.

Conover, D. O., D. A. Vanvoorhees, and A. Ehtisham. 1992. Sex-ratio selection and the evolution of environmental sex determination in laboratory populations of Menidia menidia. Evolution 46:1722-1730.

Cotton, S. and C. Wedekind. 2007a. Control of introduced species using Trojan sex chromosomes. Trends in Ecology & Evolution 22:441-443.

Cotton, S. and C. Wedekind. 2007b. Introduction of Trojan sex chromosomes to boost population growth. Journal of Theoretical Biology 249:153-161.

Cotton, S. and C. Wedekind. 2009. Population consequences of environmental sex reversal. Conservation Biology 23:196-206.

Courchamp, F., T. Clutton-Brock, and B. Grenfell. 1999. Inverse density dependence and the Allee effect. Trends in Ecology & Evolution 14:405-410.

Devlin, R. H. and Y. Nagahama. 2002. Sex determination and sex differentiation in fish: an overview of genetic, physiological, and environmental influences. Aquaculture 208:191-364.

Dobson, A. P. and A. M. Lyles. 2000. Black-footed ferret recovery. Science 288:985-988.

Ellegren, H., L. Gustafsson, and B. C. Sheldon. 1996. Sex ratio adjustment in relation to paternal attractiveness in a wild bird population. Proceedings of the National Academy of Sciences of the United States of America 93:11723-11728.

Emlen, S. T. and L. W. Oring. 1977. Ecology, sexual selection, and the evolution of mating systems. Science 197:215-223.

Felsenstein, J. 1974. The evolutionary advantage of recombination. Genetics 78:737-756.

Fisher, R. A. 1930. The genetical theory of natural selection. Clarendon Press, Oxford.

Frankham, R., J. D. Ballou, and D. A. Briscoe. 2002. Introduction to conservation genetics. Cambridge University Press, Cambridge.

Gibbons, E. F., B. S. Durrant, and J. Demarest, editors. 1995. Conservation of endangered species in captivity. State University of New York Press, Albany.

Girondot, M., H. Fouillet, and C. Pieau. 1998. Feminizing turtle embryos as a conservation tool. Conservation Biology 12:353-362.

Gomendio, M., T. H. Cluttonbrock, S. D. Albon, F. E. Guinness, and M. J. Simpson. 1990. Mammalian sex-ratios and variation in costs of rearing sons and daughters. Nature 343:261-263.

Grahn, M., A. Langesfors, and T. von Schantz. 1998. The importance of mate choice in improving viability in captive populations. Pages 341-363 in T. M. Caro, editor. Behavioral Ecology and Conservation Biology. Oxford University Press, Oxford.

Gutierrez, A., J. DeLaFuente, S. Fuentes, A. Payas, C. Ugarte, and B. Pintado. 1995. Influence of biopsy sexing and in vitro culture on losses of female mouse and bovine embryos. Animal Biotechnology 6:101-109.

Gutierrez, J. B. and J. L. Teem. 2006. A model describing the effect of sex-reversed YY fish in an established wild population: the use of a Trojan Y chromosome to cause extinction of an introduced exotic species. Journal of Theoretical Biology 241:333-341.

Hanfling, B., F. Edwards, and F. Gherardi. 2011. Invasive alien Crustacea: dispersal, establishment, impact and control. Biocontrol 56:573-595.

Harries, J. E., D. A. Sheahan, S. Jobling, P. Matthiessen, M. Neall, J. P. Sumpter, T. Taylor, and N. Zaman. 1997. Estrogenic activity in five United Kingdom rivers detected by measurement of vitellogenesis in caged male trout. Environmental Toxicology and Chemistry 16:534-542.

Hartl, D. L. 1988. A primer of population genetics, second edition. Sinauer Associates, Inc., Sunderland, Massachusetts.

Heath, D. D., J. W. Heath, C. A. Bryden, R. M. Johnson, and C. W. Fox. 2003. Rapid evolution of egg size in captive salmon. Science 299:1738-1740.

Hedrick, P. W. 2005. Genetics of populations, 3rd edition. Jones & Bartlett Publishers, Sudbury, MA, USA.

Hillis, D. M. and D. M. Green. 1990. Evolutionary changes of heterogametic sex in the phylogenetic history of amphibians. Journal of Evolutionary Biology 3:49-64.

Hurley, M. A., P. Matthiessen, and A. D. Pickering. 2004. A model for environmental sex reversal in fish. Journal of Theoretical Biology 227:159-165.

Jacob, A., G. Evanno, B. A. von Siebenthal, C. Grossen, and C. Wedekind. 2010. Effects of different mating scenarios on embryo viability in brown trout. Molecular Ecology 19:5296-5307.

Janzen, F. J. 1994. Climate change and temperature-dependent sex determination in reptiles. Proceedings of the National Academy of Sciences of the United States of America 91:7487-7490.

Janzen, F. J., J. W. Gibbons, J. L. Greene, J. B. Iverson, and J. K. Tucker. 2006. Climate change and temporal variation in nesting biology of North American turtles. Integrative and Comparative Biology 46:E69-E69.

Jobling, S. and C. R. Tyler. 2003. Endocrine disruption in wild freshwater fish. Pure and Applied Chemistry 75:2219-2234.

Jobling, S., R. Williams, A. Johnson, A. Taylor, M. Gross-Sorokin, M. Nolan, C. R. Tyler, R. van Aerle, E. Santos, and G. Brighty. 2006. Predicted exposures to steroid estrogens in UK rivers correlate with widespread sexual disruption in wild fish populations. Environmental Health Perspectives 114:32-39.

Kamel, S. J. and N. Mrosovsky. 2006. Deforestation: risk of sex ratio distortion in hawsbill sea turtles. Ecological Applications 16:923-931.

Kanaiwa, M. and Y. Harada. 2002. Genetic risk involved in stock enhancement of fish having environmental sex determination. Population Ecology 44:7-15.

Kapuscinski, A. R., K. R. Hayes, and L. Sifa, editors. 2007. Environmental risk assessment of genetically modified organisms, volume 3: transgenic fish in developing countries. CAB International Publishing, Wallingford, UK.

King, W. A., L. Picard, D. Bousquet, and A. K. Goff. 1992. Sex-dependent loss of bisected bovine morulae after culture and freezing. Journal of Reproduction and Fertility 96:453-459.

Kolar, C. S. and D. M. Lodge. 2002. Ecological predictions and risk assessment for alien fishes in North America. Science 298:1233-1236.

Krackow, S. 1995. The developmental asynchrony hypothesis for sex-ratio manipulation. Journal of Theoretical Biology 176:273-280.

Lande, R. 1998. Anthropogenic, ecological and genetic factors in extinction. Pages 29-51 in G. M. Mace, A. Balmford, and J. R. Ginsberg, editors. Conservation in a changing world. Cambridge University Press, Cambridge.

Lanza, R. P., B. L. Dresser, and P. Damiani. 2000. Cloning Noah's ark. Scientific American 283:84-89.

Larsson, D. G., H. Hallman, and L. Förlin. 2000. More male fish embryos near a pulp mill. Environmental Toxicology and Chemistry 19:2911-2917.

Lenz, T. L., A. Jacob, and C. Wedekind. 2007. Manipulating sex ratio to increase population growth: the example of the Lesser Kestrel. Animal Conservation 10:236-244.

Louda, S. M. and P. Stiling. 2004. The double-edged sword of biological control in conservation and restoration. Conservation Biology 18:50-53.

Magerhans, A., A. Müller-Belecke, and G. Hörstgen-Schwark. 2009. Effect of rearing temperatures post hatching on sex ratios of rainbow trout (*Oncorhynchus mykiss*) populations. Aquaculture 294:25-29.

Marealle, W. N., F. Fossoy, T. Holmern, B. G. Stokke, and E. Roskaft. 2010. Does illegal hunting skew Serengeti wildlife sex ratios? Wildlife Biology 16:419-429.

Milinski, M. and T. C. M. Bakker. 1992. Costs influence sequential mate choice in sticklebacks, *Gasterosteus aculeatus*. Proceedings of the Royal Society of London Series B-Biological Sciences 250:229-233.

Muir, W. M. and R. D. Howard. 2004. Characterization of environmental risk of genetically engineered (GE) organisms and their potential to control exotic invasive species. Aquatic Sciences 66:414-420.

Muller, H. J. 1932. Some genetic aspects of sex. American Naturalist 66:118-138.

Myers, J. H., D. Simberloff, A. M. Kuris, and J. R. Carey. 2000. Eradication revisited: dealing with exotic species. Trends in Ecology & Evolution 15:316-320.

Nagler, J. J., J. Bouma, G. H. Thorgaard, and D. D. Dauble. 2001. High incidence of a male-specific genetic marker in phenotypic female Chinook salmon from the Columbia River. Environmental Health Perspectives 109:67-69.

Nomura, T. 2002. Effective size of populations with unequal sex ratio and variation in mating success. Journal of Animal Breeding and Genetics 119:297-310.

Olsen, J. B., S. J. Miller, K. Harper, J. J. Nagler, and J. K. Wenburg. 2006. Contrasting sex ratios in juvenile and adult chinook salmon *Oncorhynchus tshawytscha* (Walbaum) from south-west Alaska: sex reversal or differential survival? Journal of Fish Biology 69:140-144.

Ospina-Alvarez, N. and F. Piferrer. 2008. Temperature-dependent sex determination in fish revisited: prevalence, a single sex ratio response pattern, and possible effects of climate change. PLoS ONE 3:e2837.

Palmer, A. R. 2000. Quasi-replication and the contract of error: lessons from sex ratios, heritabilities and fluctuating asymmetry. Annual Review of Ecology and Systematics 31:441–480.

Pandian, T. J. and S. G. Sheela. 1995. Hormonal induction of sex reversal in fish. Aquaculture 138:1-22.

Parks, L. G., C. S. Lambright, E. F. Orlando, L. J. Guillette, G. T. Ankley, and L. E. Gray. 2001. Masculinization of female Mosquitofish in a kraft mill effluent-contaminated Fenholloway River water is associated with androgen receptor agonist activity. Toxicological Sciences 62:257-267.

Penáz, M., Z. Svbodová, V. Barus, M. Prokes, and J. Drastichova. 2005. Endocrine disruption in a barbel, Barbus barbus population from the River Jihlava. Journal of Applied Ichthyology 21:420-428.

Perret, M. 1996. Manipulation of sex ratio at birth by urinary cues in a prosimian primate. Behavioral Ecology and Sociobiology 38:259-266.

Perrin, N. 2009. Sex reversal: a fountain of youth for sex chromosomes? Evolution 63:3043-3049.

Piferrer, F. 2001. Endocrine sex control strategies for the feminization of teleost fish. Aquaculture 197:229-281.

Poulin, R., R. A. Paterson, C. R. Townsend, D. M. Tompkins, and D. W. Kelly. 2011. Biological invasions and the dynamics of endemic diseases in freshwater ecosystems. Freshwater Biology 56:676-688.

Rankin, D. J. and H. Kokko. 2007. Do males matter? The role of males in population dynamics. Oikos 116:335-348.

Real, L. A. 1991. Search theory and mate choice .2. mutual interaction, assortative mating, and equilibrium variation in male and female fitness. American Naturalist 138:901-917.

Rice, W. R., S. Gavrilets, and U. Friberg. 2008. Sexually antagonistic "zygotic drive" of the sex chromosomes. Plos Genetics 4.

Robertson, B. C., G. P. Elliott, D. K. Eason, M. N. Clout, and N. J. Gemmell. 2006. Sex allocation theory aids species conservation. Biology Letters 2:229-231.

Ryman, N. and L. Laikre. 1991. Effects of supportive breeding on the genetically effective population-size. Conservation Biology 5:325-329.

Saino, N., H. Ellegren, and A. P. Møller. 1999. No evidence for adjustment of sex allocation in relation to paternal ornamentation and paternity in barn swallows. Molecular Ecology 8:399-406.

Schliekelman, P., S. Ellner, and F. Gould. 2005. Pest control by genetic manipulation of sex ratio. Journal of Economic Entomology 98:18-34.

Schliekelman, P. and F. Gould. 2000. Pest control by the release of insects carrying a female-killing allele on multiple loci. Journal of Economic Entomology 93:1566-1579.

Solari, A. J. 1994. Sex chromosomes and sex determination in vertebrates. CRC Press, Boca Raton, Florida.

Stelkens, R. B. and C. Wedekind. 2010. Environmental sex reversal, Trojan sex genes, and sex ratio adjustment: conditions and population consequences. Molecular Ecology 19:627-646.

Stephens, P. A. and W. J. Sutherland. 1999. Consequences of the Allee effect for behaviour, ecology and conservation. Trends in Ecology & Evolution 14:401-405.

Sumpter, J. P. and S. Jobling. 1995. Vitellogenesis as a biomarker for estrogenic contamination of the aquatic environment. Environmental Health Perspectives 103:173-178.

Tella, J. L. 2001. Sex-ratio theory in conservation biology. Trends in Ecology & Evolution 16:76-77.

Thailayil, J., K. Magnusson, H. C. J. Godfray, A. Crisanti, and F. Catteruccia. 2011. Spermless males elicit large-scale female responses to mating in the malaria mosquito Anopheles gambiae. Proceedings of the National Academy of Sciences of the United States of America 108:13677-13681.

Tinbergen, N. 1963. On aims and methods of ethology. Zeitschrift fur Tierpsychologie 20:410-433.

Trivers, R. L. and D. E. Willard. 1973. Natural selection of parental ability to vary the sex ratio of offspring. Science 179:90-91.

Tryjanowski, P., T. H. Sparks, R. Kamieniarz, and M. Panek. 2009. The relationship between hunting methods and sex, age and body weight in a non-trophy animal, the red fox. Wildlife Research 36:106-109.

Valenzuela, N. and V. A. Lance, editors. 2004. Temperature-dependent sex determination in vertebrates. Smithsonian Institution, Washington.

Vos, J. G., E. Dybing, H. A. Greim, O. Ladefoged, C. Lambre, J. V. Tarazona, I. Brandt, and A. D. Vethaak. 2000. Health effects of endocrine-disrupting chemicals on wildlife, with special reference to the European situation. Critical Reviews in Toxicology 30:71-133.

Wallace, H., G. M. I. Badawy, and B. M. N. Wallace. 1999. Amphibian sex determination and sex reversal. CMLS Cellular and Molecular Life Sciences 55:901-909.

Wedekind, C. 2002a. Manipulating sex ratios for conservation: short-term risks and long-term benefits. Animal Conservation 5:13-20.

Wedekind, C. 2002b. Sexual selection and life-history decisions: implications for supportive breeding and the management of captive populations. Conservation Biology 16:1204-1211.

Wedekind, C. 2010. Searching for sex-reversals to explain population demography and the evolution of sex chromosomes. Molecular Ecology 19:1760-1762.

Wedekind, C. and C. Küng. 2010. Shift of spawning season and effects of climate warming on developmental stages of a grayling (Salmonidae). Conservation Biology 24:1418-1423.

Wedekind, C., R. Müller, and H. Spicher. 2001. Potential genetic benefits of mate selection in whitefish. Journal of Evolutionary Biology 14:980-986.

Wedekind, C., G. Rudolfsen, A. Jacob, D. Urbach, and R. Müller. 2007. The genetic consequences of hatchery-induced sperm competition in a salmonid. Biological Conservation 137:180-188.

Westerdahl, H., S. Bensch, B. Hansson, D. Hasselquist, and T. VonSchantz. 1997. Sex ratio variation among broods of great reed warblers Acrocephalus arundinaceus. Molecular Ecology 6:543-548.

Williamson, K. S., R. Phillips, and B. May. 2008. Characterization of a chromosomal rearrangement responsible for producing "Apparent" XY-female fall-run Chinook salmon in California. Journal of Heredity 99:483-490.

Young, A. G. and G. M. Clarke, editors. 2000. Genetics, demography and viability of fragmented populations. Cambridge University Press, Cambridge UK.

Permissions

The contributors of this book come from diverse backgrounds, making this book a truly international effort. This book will bring forth new frontiers with its revolutionizing research information and detailed analysis of the nascent developments around the world.

We would like to thank Tony Povilitis, for lending his expertise to make the book truly unique. He has played a crucial role in the development of this book. Without his invaluable contribution this book wouldn't have been possible. He has made vital efforts to compile up to date information on the varied aspects of this subject to make this book a valuable addition to the collection of many professionals and students.

This book was conceptualized with the vision of imparting up-to-date information and advanced data in this field. To ensure the same, a matchless editorial board was set up. Every individual on the board went through rigorous rounds of assessment to prove their worth. After which they invested a large part of their time researching and compiling the most relevant data for our readers. Conferences and sessions were held from time to time between the editorial board and the contributing authors to present the data in the most comprehensible form. The editorial team has worked tirelessly to provide valuable and valid information to help people across the globe.

Every chapter published in this book has been scrutinized by our experts. Their significance has been extensively debated. The topics covered herein carry significant findings which will fuel the growth of the discipline. They may even be implemented as practical applications or may be referred to as a beginning point for another development. Chapters in this book were first published by InTech; hereby published with permission under the Creative Commons Attribution License or equivalent.

The editorial board has been involved in producing this book since its inception. They have spent rigorous hours researching and exploring the diverse topics which have resulted in the successful publishing of this book. They have passed on their knowledge of decades through this book. To expedite this challenging task, the publisher supported the team at every step. A small team of assistant editors was also appointed to further simplify the editing procedure and attain best results for the readers.

Our editorial team has been hand-picked from every corner of the world. Their multi-ethnicity adds dynamic inputs to the discussions which result in innovative outcomes. These outcomes are then further discussed with the researchers and contributors who give their valuable feedback and opinion regarding the same. The feedback is then collaborated with the researches and they are edited in a comprehensive manner to aid the understanding of the subject.

Apart from the editorial board, the designing team has also invested a significant amount of their time in understanding the subject and creating the most relevant covers. They scrutinized every image to scout for the most suitable representation of the subject and create an appropriate cover for the book.

The publishing team has been involved in this book since its early stages. They were actively engaged in every process, be it collecting the data, connecting with the contributors or procuring relevant information. The team has been an ardent support to the editorial, designing and production team. Their endless efforts to recruit the best for this project, has resulted in the accomplishment of this book. They are a veteran in the field of academics and their pool of knowledge is as vast as their experience in printing. Their expertise and guidance has proved useful at every step. Their uncompromising quality standards have made this book an exceptional effort. Their encouragement from time to time has been an inspiration for everyone.

The publisher and the editorial board hope that this book will prove to be a valuable piece of knowledge for researchers, students, practitioners and scholars across the globe.

List of Contributors

S. Naghmouchi, M. L. Khouja, A. Khaldi and M. N. Rejeb
Institute of Research in Rural Engineering, Water and Forestry (INRGREF), Tunisia, Tunis

S. Zgoulli and P. Thonart
Wallon Centre of Industrial Biology (CWBI), University of Liege Sart Tilman, Liège, Belgique

M. Boussaid
National Institute of Applied Science to Technology (INSAT Tunis), Centre Urbain Nord, Tunisia, Tunis

Brandon P. Anthony and Diane A. Matar
Environmental Sciences & Policy Department, Central European University, Budapest, Hungary

Hiroyuki Iketani
Institute of Fruit Science, National Agriculture and Food Research Organization, Food Resources Education and Research Center, Japan

Hironori Katayama
Graduate School of Agricultural Science, Kobe University, Japan

Tatyanna Mariúcha de Araújo Pantoja
Museu Paraense Emílio Goeldi MPEG, Universidade Federal do Pará UFPA, Brazil

Fernando César Weber Rosas and Vera Maria Ferreira Da Silva
Departamento de Biologia Aquática e Limnologia, Divisão de Ictiologia e Mamíferos Aquáticos, Instituto Nacional de Pesquisas da Amazônia INPA, Brazil

Ângela Maria Fernandes Dos Santos
Instituto de Hematologia e Hemoterapia do Amazonas – HEMOAM, Brazil

Stephen M. Awoyemi
Tropical Conservancy, Nigeria

Amy Gambrill
International Resources Group, USA

Alison Ormsby
Eckerd College, USA

Dhaval Vyas
Alverado Way, Decatur, Georgia, USA

Claus Wedekind
Department of Ecology and Evolution, Biophore, University of Lausanne, Lausanne, Switzerland

Printed in the USA
CPSIA information can be obtained
at www.ICGtesting.com
JSHW011326221024
72173JS00003B/72